SOLID-STATE CIRCUIT TROUBLESHOOTING GUIDE

by art margolis

TAB BOOKS
Blue Ridge Summit, Pa. 17214

FIRST EDITION

FIRST PRINTING—JUNE 1972

Printed in the United States
of America

Hardbound Edition: International Standard Book No. 0-8306-2607-7

Paperbound Edition: International Standard Book No. 0-8306-1607-1

Library of Congress Card Number: 72-81362

PREFACE

In comparison to vacuum tube circuits, solid-state circuits are tiny. As a result, great care must be taken during soldering and testing. A bad move can decimate the tiny components or a too high test voltage can blow a transistor. However, the technician adapts quickly to the miniaturization.

Since solid-state devices simply replace vacuum tubes, the general layout of electronic instruments, from a block diagram point of view, are essentially the same in both forms. Even the resistive, capacitance and inductive elements are quite the same. The big difference is in the solid-state devices themselves and the source supply voltages needed to power them.

The troubleshooter must develop thinking patterns that relate to solid state. These patterns evolve with improvement in his ability to look at a solid-state schematic and immediately envision the way the electrons are flowing through the circuit and the way diodes, transistors and FETs are affecting those electrons. Then his mind tells him what pieces of test equipment to reach for and where to begin looking for the cause of trouble. From this analysis he is directed on a pathway that, by the process of elimination, enables him to isolate the defective parts or connections. Once the part is identified the rest is easy and mechanical, even though it's miniaturized.

In the first four chapters I discuss the various solid-state devices from my point of view as a troubleshooter. The rest of the chapters are discussions of how these devices work in typical circuits. At various sections of the book there are schematics of actual solid-state circuits and suggestions on how to align and service them. Of particular interest are schematics and "service thinking" charts developed by Motorola and General Electric. They exemplify the best technique of taking a schematic, analyzing it carefully and then following a pathway until the defect is encountered.

I would like to give my sincerest thanks to the following people and companies for their cooperation in helping me with the book: General Electric: R. C. Hannum, Supervisor-Training & Technical Publications; Motorola Consumer Products: El Mueller, Manager, Technical Training; RCA: W. H. Fulroth, Manager, Technical Publications; Sylvania: E. M. Nanni, Technical Publications Manager; Zenith Sales Company: E. Kob, Assistant National Service Manager. I'd also like to thank my wife, Lea, and my typist, Emily Burtis, for reading my handwriting. Last but not least, credit for the cover photo goes to my son Denny.

Art Margolis

CONTENTS

PREFACE 3

1 SOLID-STATE DEVICE "CONSTRUCTION" 7
Covalent Bonds—Making the P and N Materials—N Material—P Material—PN Junction—Junction Bias—Testing the Junction

2 DIODES & BIPOLAR TRANSISTORS 24
Detector Diodes—Diode Capacitance—Zener Diodes—Varicap Diodes—Bipolar Transistors—Testing Diodes—Testing Transistors

3 FETs 45
FETs—JFETs—IGFETs & MOSFETs—Testing FETs

4 ICs & SCRs 59
Integrated Circuits—Testing ICs—Silicon Controlled Rectifiers

5 RF AMPLIFIER CIRCUITS 70
Bipolar RF Amplifiers—RF AGC—Field-Effect Transistor RF Amplifiers—Testing RF Circuits—TV RF Alignment

6 IF AMPLIFIER CIRCUITS 88
Bandpass—Transistor IF Stage—IF Transformer—Amplifier—Stabilization—Coupling Components—IF AGC—Testing IF Circuits—Alignment—TV—Marker Injection

7 AUDIO & VIDEO AMPLIFIERS 110
Audio & Video Amplifiers—Audio Amplifiers—Class A and Class B Amplifiers—Voltage Amplifier—Power

Amplifier—Video Amplifiers—Testing Audio and Video Circuits—Troubleshooting Charts

8 POWER AMPLIFIERS 128
Pushpull Power Amplifiers—Transformerless Pushpull Outputs—Vertical Output Circuits—Horizontal Output Circuits—Transmitter RF Power Amplifiers—Testing Power Amplifier Circuits—Troubleshooting Charts

9 OSCILLATORS 145
Oscillator Multipliers—LC Oscillators—RC Type Oscillators—Testing Oscillator Circuits

10 CONVERTER CIRCUITS 156
Conversion Loss—Transistor Converter—Transistor Mixer—JFET Mixer—Dual-Gate Mixer—Diode Mixer—Direct-Conversion Circuits—Testing Converter Circuits—Transistor Coupling Transformers

11 DETECTORS 167
Time Constant—FM Detectors—Ratio Detector

12 AUTOMATIC ADJUSTMENT CIRCUITS 178
AGC—AFC—Varactor Diodes (Varicaps)— Trouble-shooting Charts

13 REMOTE CONTROL CIRCUITS 192
RF Remotes—Transducer Remotes—Typical Transmitters—Typical Receivers

14 SEPARATOR CIRCUITS 203
Sync Separation—Separator Circuit—Noise Gate—Sync Pulse Separation—Color Sync Separator—Testing Separation Circuits

15 POWER SUPPLIES 213
Half-Wave—Full-Wave Bridge—Active Power Filters—High-Voltage Power Supplies—Voltage Multiplication

INDEX 221

CHAPTER 1

Solid-State Device "Construction"

Solid-state P materials and N materials are the building blocks of all semiconductor devices. A simple diode is nothing more than one piece of P material bonded to a piece of N material (Fig. 1-1). A PNP transistor is a diode with another piece of P material bonded on the other side of the N material, making a PNP sandwich. An NPN transistor is a diode with an extra piece of N material, making an NPN sandwich. An FET is another sandwich, but with the input and output connections both attached to the inside of the sandwich or the channel. A silicon controlled rectifier is two P-N diodes bonded together so it takes on the characteristics of a rectifier and a transistor.

As the name implies, semiconductor material, whether it is P or N, is not a conductor or an insulator, but something in between. The names conductor, semiconductor and insulator refer to the way electrons move through the material. A conductor requires small amounts of voltage pressure to push large amounts of electrons through the material. A semiconductor requires a moderate amount of voltage pressure to push a moderate amount of electrons through the material. An insulator requires great amounts of voltage to push minute amounts of electrons through the material.

COVALENT BONDS

The term, covalent bonds, is important since it describes exactly the material state of the two most common semiconductor substances—silicon and germanium. Under a microscope, a silicon atom looks like a little solar system. It has a nucleus like the sun and 14 electrons traveling around the nucleus like the planets (Fig. 1-2). In the outermost orbit

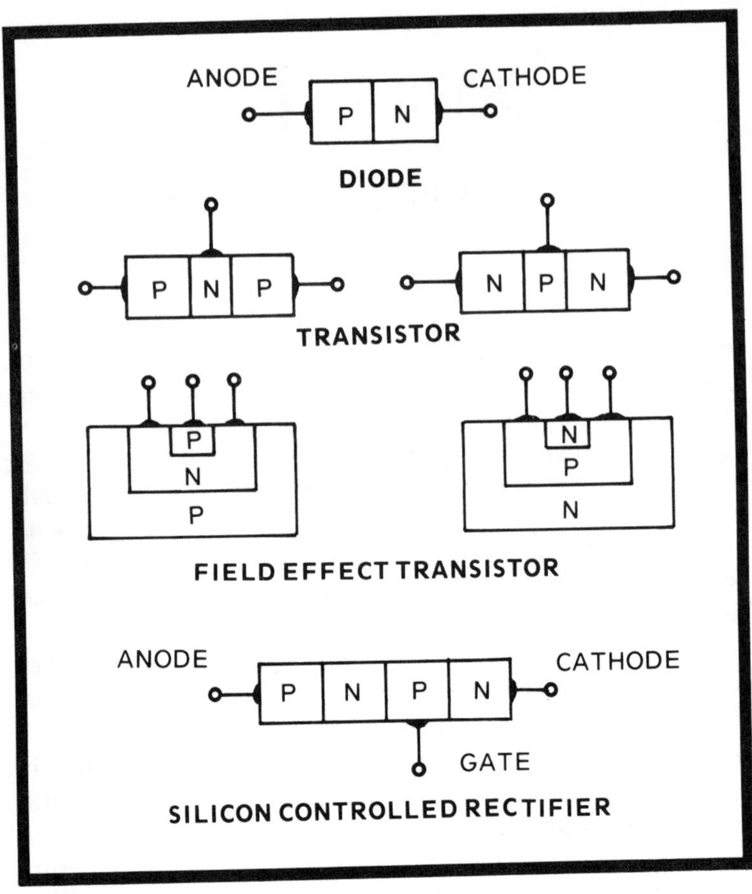

Fig. 1-1. All so called "solid-state" devices are made up of the basic N and P building blocks.

there are four electrons. These electrons are seeking stability which is called an "inert state." All atomic structures seek an inert state. The inert state in a silicon atom is reached when the number of electrons in the outer orbit of an atom is increased from four to eight (Fig. 1-3). Why do eight electrons produce inertness? I don't know, it's a law of the universe.

A silicon atom has an easy time reaching inertness. It simply combines with another silicon atom and they share the electrons in their outer orbit. The outer orbit is called the valence shell and the electrons in it are the valence electrons. Regarding troubleshooting semiconductor materials, we are

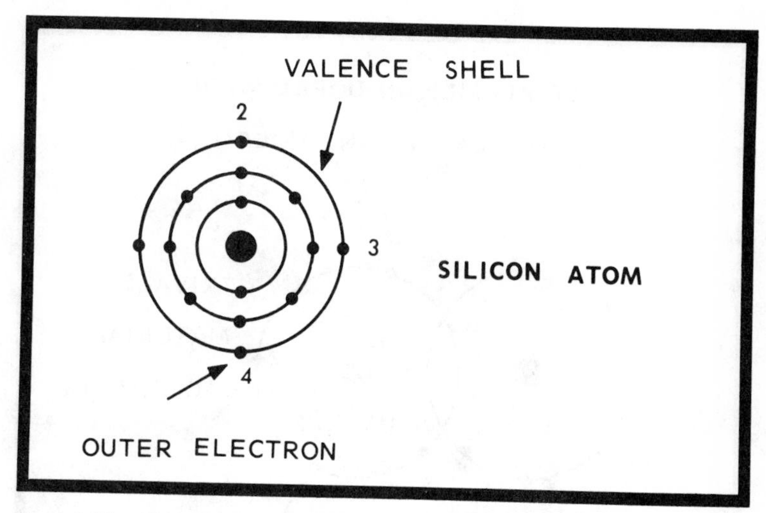

Fig. 1-2. The silicon atom has a valence shell containing four electrons in orbit. These are the current makers.

interested only in the valence electrons; the rest of the atom is of little interest.

When two silicon atoms combine their valence electrons, they become a silicon molecule and are said to be "covalently bonded." They have become inert as desired and are now a good insulator. That's because the 8-electron configuration has such tight covalent bonding that it is very difficult to free an electron for conduction purposes. In this state the silicon is

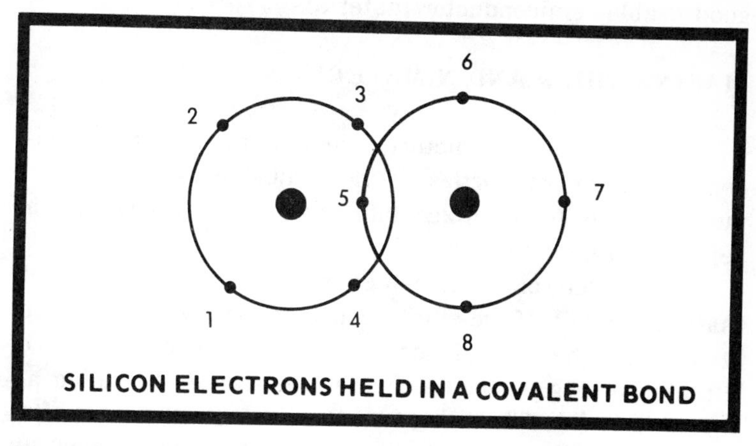

Fig. 1-3. The electrons in the valence shell form a stable bond in configurations of eight.

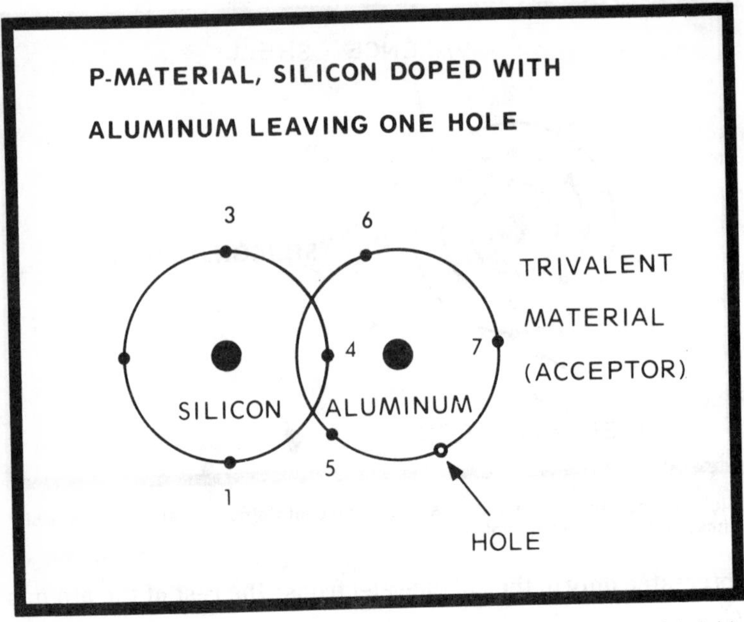

Fig. 1-4. A piece of P material is created by removing one of the eight stable electrons, making a hole.

not a semiconductor but an insulator. It has also attained a crystal structure. A crystal structure is a "collection of atoms oriented with respect to each other in a definite orderly manner." The crystal structure, however, can be made into a good usable semiconductor material.

MAKING THE P AND N MATERIALS

The crystalline structure, when pure, has no special semiconductor properties. Other substances, called impurities, have to be mixed into the crystal to produce the separate properties.

A pure piece of silicon can be made into a piece of P or N material at will. If the silicon can be doped with a trivalent atom (one with three valence electrons), it becomes a piece of P material (Fig. 1-4). The trivalent materials are called **acceptor** materials. That's because the 3-valent substance, after it combines with a 4-valent atom of silicon, forms a 7-electron valence shell. Since the inert state is eight, the 7-valence material tries to go to eight and is always ready to "accept" another electron. Typical trivalent materials are aluminum, gallium and boron.

When the pure silicon can be doped with a pentavalent atom, one with five valence electrons, it becomes a piece of N material (Fig. 1-5). The pentavalent materials are called **donor** materials. That's because the 5-valent substance, after it combines with a 4-valent atom of silicon, forms a 9-valent outer shell. Since the inert state is eight, the 9-valence material tries to go back to eight and is always ready to "donate" its extra electron. Typical pentavalent materials are arsenic, antimony and phosphorous.

It doesn't take much to dope the silicon. One atom of doping material to millions of atoms of silicon does the trick. Germanium is another material that lends itself to doping, thus forming pieces of P and N material. While the characteristics of germanium are quite different than silicon, from the P and N point of view they are remarkably similar. Therefore, this discussion about the silicon transformation

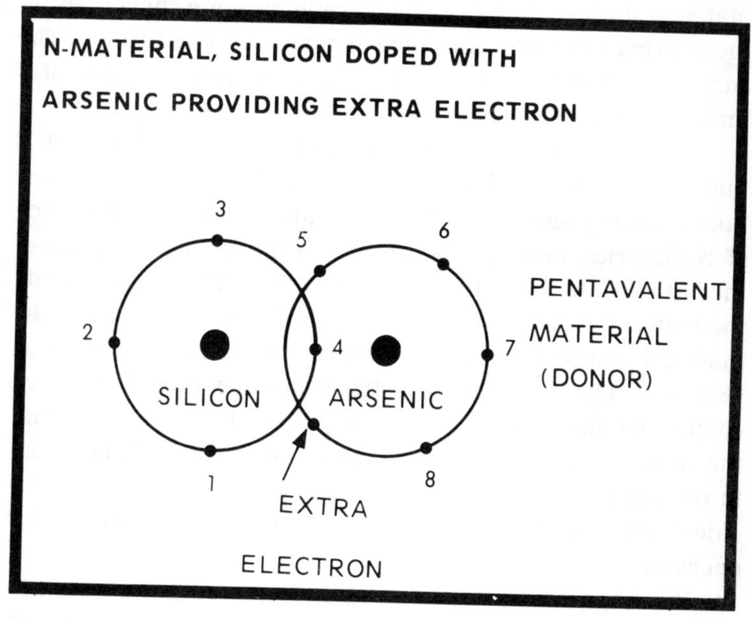

Fig. 1-5. A piece of N material is created by adding one to the eight stable electrons, making an excess of electrons.

from an inert crystalline substance is also pertinent to changing germanium from an inert to a semiconductor P or N material.

N MATERIAL

When a piece of crystal silicon or germanium is doped with some donor material, or a pentavalent material like arsenic, the crystal and the donor go into a covalent bond. These combined materials, spaced out through the total crystalline material, have nine electrons in the outer shell. However, there is only "seating room" for eight. The ninth electron has no permanent place and roams loose around the outer shell.

In a piece of N material there are millions of roaming electrons, which are negative charges, revolving around the parent atom. Any little bits of energy, such as light, heat or current, easily frees the extra electron from the parent atom and the electron can roam haphazardly through the piece of N material. In fact, there is considerable random movement of these extra electrons about the N-material at all times. The only way to halt the electron roaming is to place the piece of N material in a deep freeze.

Roaming electrons are current carriers. In a piece of N material the electrons are called the **majority** carriers. If you take a battery and attach the terminals to two ends of a piece of N material, electrons will travel from the negative end of the battery into the N material. The roaming electrons inside the material will be repelled and head for the other end of the material where the positive end of the battery is attached (Fig. 1-6). The N material electrons will flow into the positive connection and be replaced in the material by electrons from the battery. A current will flow from one end of the N material to the other as long as the battery is attached (Fig. 1-6). Incidentally, this is the principle by which an N-channel FET operates (see Chapter 3).

It is important in solid-state troubleshooting to constantly see, in the minds eye, the electron flow through the semiconductor materials. It is also important in vacuum tube troubleshooting, but the electrons in a vacuum tube flow only in one direction—from cathode to plate. In solid-state devices the electrons can be flowing from emitter to base and collector (NPN) or they can be flowing from collector and base to emitter (PNP). In FETs, the electrons can be flowing from source to drain (N channel) or from drain to source (P channel). It is important that you "see" at a glance the actual electron flow as you peruse a schematic or a chassis.

Electrons flow from an excess of electrons, or a negative charge, to a scarcity of electrons, or a positive charge. That is where the confusion enters the scene. An excess is not usually thought of as negative, and a scarcity is not usually thought of as positive. It is an unfortunate use of terms.

It is a good idea to practice thinking of electrons flowing toward a power supply positive voltage, rather than electricity coming from the power supply. In vacuum tube thinking, the technician "sees" voltage coming from the power supply. In solid-state thinking the technician must "see" electrons coming out of the chassis ground and flowing towards the positive voltage potential, or a point of electron scarcity. The

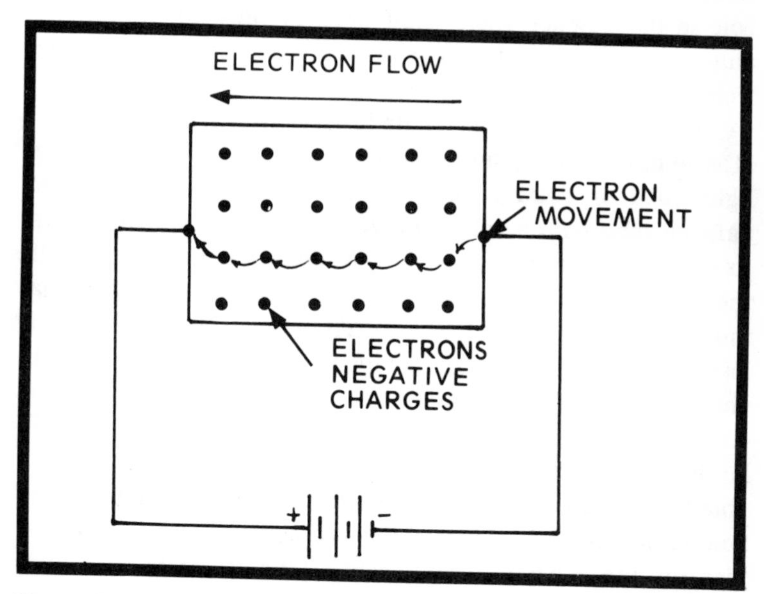

Fig. 1-6. Electrons from an outside power source can cause a current path in N material by repelling the extra electrons.

only way electrons can flow toward chassis ground is if the power supply point is negative, or has excess electrons. Then, electrons can flow from B minus toward chassis ground.

P MATERIAL

When considering N material, analogies are easily understood, since the flow of electrons is quite similar to the way electrons flow in vacuum tube circuits. The N material by itself is almost a conductor, just like any other conductor, and it passes electrons like a conductor. This is not true with P material. P material brings into play an entirely new concept. Rather than considering only electron flow, you must also consider **hole** flow.

When P material is made, the same type of silicon or germanium is doped with a trivalent substance such as aluminum. The aluminum is introduced into the crystalline structure and combines with the crystals to form covalent bonds. Since aluminum has only three electrons in its valence orbit, the crystal-aluminum substance formed has only seven electrons, one shy of the eight required for inertness, leaving a hole in the valence shell. That hole enables all the crystal-aluminum atomic pairs to be able to "accept" another electron.

The fact that there is a hole in the valence shell keeps the atomic pair from becoming inert. Any bits of energy such as light, heat or voltage potential can cause electrons in the valence shell with a hole to move. If an electron moves from its position in the shell to the hole as energy activates it, it fills the hole with a negative charge. The position where that moving electron originated loses its negative charge. Since it lost a negative charge, it becomes more positive or positively charged. Therefore, a hole is a positive charge.

An electron not only is able to move about in its own orbit, but it can jump from one covalent bond to another, if there is a hole for it to go to. That way, as the electrons jump from one hole to another, the holes, which are actual positive charges, move in the opposite direction from the electrons.

A useful analogy of hole movement is to think of five boxes numbered one through five in a line. Number one is empty or has a hole, while the other four are full. If you take the contents of number two and throw it into the empty number one, then number one fills up and the hole has moved to number two. If you keep doing that in numerical sequence, the hole will move to number three next, then four and five. A piece of P-material, full of holes, has a lot of random hole movement. Consequently, the hole is the majority carrier.

Of course, there is electron movement in the P material. The electrons still flow from an excess or negative to a scarcity or positive voltage charge. As the electrons move about, the positive charges or holes flow in the other direction. They flow from an excess of holes (a positive charge) to a scarcity of holes (a negative charge). A hole doesn't exist as a majority carrier in a piece of N material or for that matter in any conductor. It exists only in a piece of P material simply because of the unique combination of the silicon or germanium crystal structure with a trivalent substance.

Even though there are negative and positive charges in spots on a piece of N or P material, there is no total charge on the material. It is ready for conduction, but the electrons and holes are held tightly in the crystal and the net charge of the material is zero.

A hole is a specific positive charge, just as an electron is a specific negative charge. Both the hole and the electron are affected by voltage potential. A hole is attracted by a negative charge and repelled by a positive charge. An electron is attracted by a positive charge and repelled by a negative charge. Holes abound in P material while electrons abound in N material. The holes and the excess of electrons are the medium by which current flow takes place in the crystalline structure. Without them the silicon or germanium is not a semiconductor but an insulator.

If a battery is attached to a piece of P material in the manner in which it was attached to the N material, the electrons are attracted to the positive terminal just like a piece of N material (Fig. 1-7). However, since there is a majority of holes, as the electrons move toward the positive terminal, they jump from one hole to another as they go, rather than hopping from one full valence shell to another. As a result the holes are

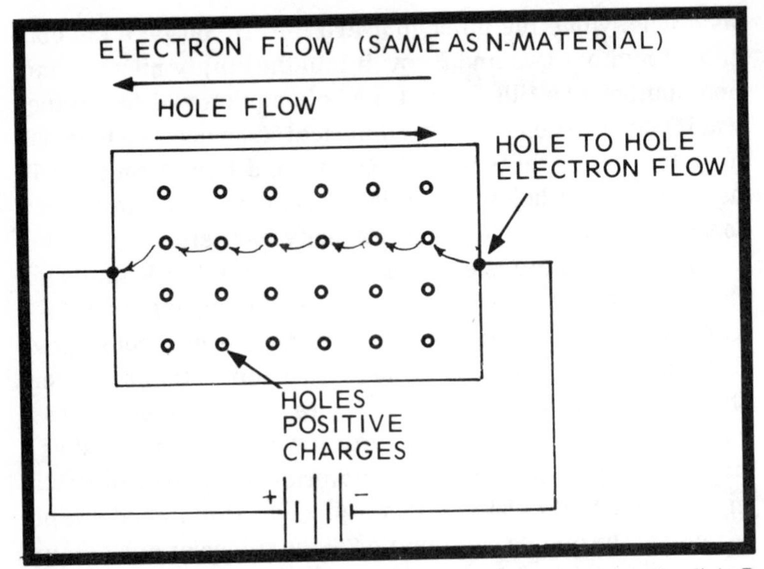

Fig. 1-7. Electrons from an outside power source can cause a current path in P material by traveling from hole to hole.

moved toward the negative terminal. In other words, holes are positive charges and move as they are created to move.

Actually, electrons flow in the same direction in both the P and N materials. Electrons enter the P and N materials from the negative terminal and depart at the positive terminal. The big difference in the two materials is the presence of the holes in the P material. In the N material the electrons, being the majority carrier, flow through a bunch of other electrons. In the P material the holes, being the majority carrier, flow toward the negative terminal at the same time that the electrons are entering from the negative terminal. To put it another way, there is a one-way traffic pattern in the N material and two-way traffic in the P material. The one-way traffic is all electrons while the two-way traffic consists of electrons going one way and holes going the other way. The importance of this concept is emphasized in the following P-N junction discussion.

Minority Carriers

All the talk about majority carriers brings up the question: What about minority carriers? In an N material the minority carrier is composed of holes and in P material it's electrons. Even though a piece of N material is doped with pentavalent substances, energy is absorbed in the material in such a way as to free electrons from valence shells and leave a hole. There isn't too much of this, but there is a bit. The hole is then available in the N material, just as it is in the P material, to absorb an electron from another valence shell and thus cause electron flow. Since there are very few holes in N material, in comparison to extra electrons, the holes are obviously the minority carrier in N material, while the electrons are the majority carrier.

The same type of thing also occurs in P material. There are enough bits of energy around to free electrons from the valence shell so they may roam through the material. The electron movement constitutes electron flow. Since there are so few free electrons in P material, they are the minority carriers while the holes are the majority carriers.

To sum up the material discussion, P material has an acceptor trivalent doping that creates holes as the majority carrier and electrons as the minority carrier. N material gets a donor pentavalent doping that makes the electron the majority carrier while the holes is the minority carrier. Try to memorize this paragraph. It will help to give a clear understanding of the basic building blocks in a solid-state device.

PN JUNCTION

The PN junction is the first device suggested by the existence of pieces of P material and N material. What would we create if we place one piece of P and one piece of N together? Just placing them together does nothing. However, if a permanent junction is formed at the point they touch, such as is produced by melting or "growing" the two pieces together, a solid-state diode is formed (Fig. 1-8).

The diode, complete with cathode and anode, is formed due to a phenomenon that occurs at the permanent junction. At this front line, roaming electrons are suddenly confronted with a bunch of movable holes. Since the electrons are negative charges and the holes are positive charges, the two different static charges are attracted to each other.

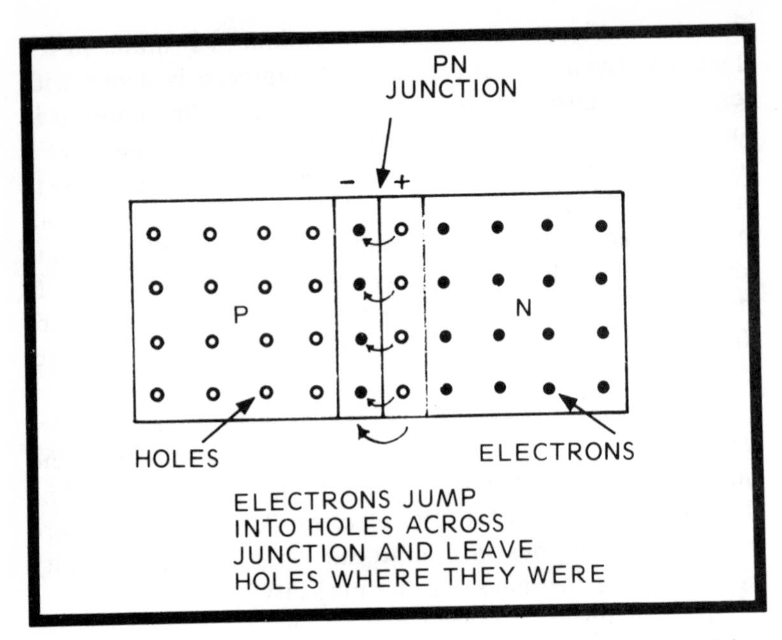

Fig. 1-8. When a piece of P material is bonded to a piece of N material, a "junction" is formed with the polarity indicated.

At the junction the electrons from the N piece jump into holes in the P piece. Every electron that jumps into a hole causes the atomic structure it came from to lose its charge and causes the structure it jumps into to lose its charge. After a group of electrons has crossed the junction and combined with P material across the junction, the next electrons up encounter a crystal area that no longer has a positive charge. Since the crystal structure is locked tight into place, the crystals present a barrier to any further movement of electrons from N to P.

However, a few more electrons do manage to cross over, but can't get any further. Also, as the electrons leave the N material they leave holes. As a result, on the P side of the junction a tiny negative charge builds up. On the N side of the junction there is a tiny positive charge. An actual voltage potential (about 0.1v) then exists at the junction. Really, it's a tiny battery from the negative to the positive. The charged area is called the **barrier** region. It also has other names like space charge region and depletion region.

This PN junction phenomenon is the basis for all diode and ordinary transistor action. It is vital that the technician understand it well. Every circuit analysis should be made with this idea in the back of the mind.

JUNCTION BIAS

If a battery or DC from a power supply is applied across a PN junction, the barrier region is affected. The barrier can become larger if a reverse bias is applied or narrower when a forward bias is applied. If the negative terminal of a battery is attached to the P material and the positive terminal is attached to the N material, the holes in the P are attracted to the battery negative end as the electrons in the N are attracted to the battery positive (Fig. 1-9).

Many electrons that were near the junction are attracted away from the junction toward the battery. Also, holes that were near the junction are attracted away toward the battery. This lessening of the number of charges near the junction widens the barrier, and the electrons that previously wanted to cross the junction and cause current flow cannot do so. The charge across the junction gets bigger and as soon as the

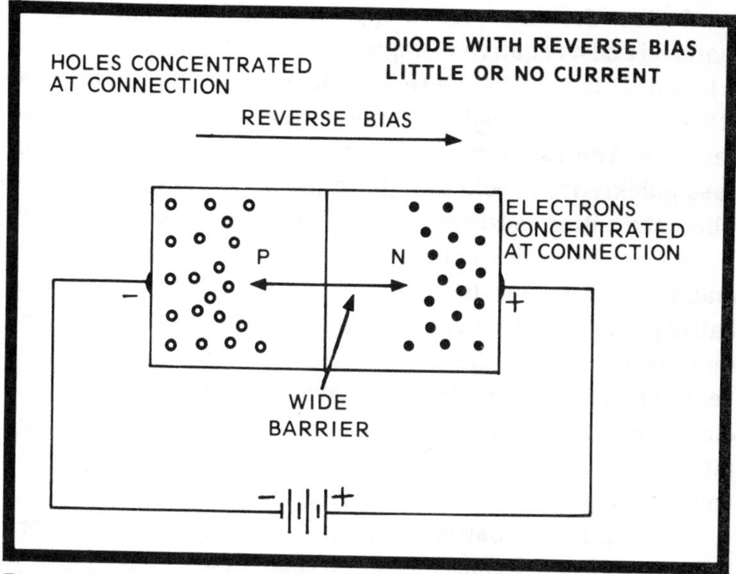

Fig. 1-9. If a PN junction is reverse biased, the junction's width is increased, making it a very high resistance.

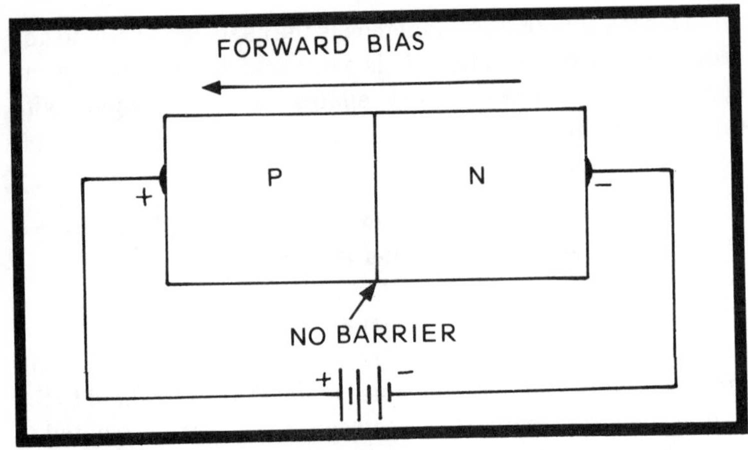

Fig. 1-10. If a PN junction has a forward bias, the junction is drastically narrowed, making it a very low resistance.

charge equals the battery voltage, all current flow comes to a stop. This is reverse bias. The PN junction develops a condition of no current flow under these conditions. There is a limit to the amount of reverse bias that can be applied. If there is too much reverse bias the crystals can fall apart and a heavy current will start to flow.

On the other hand, if you apply a positive voltage to the P material and a negative voltage to the N material, the junction is forward biased (Fig. 1-10). The electrons at the junction are attracted to the positive voltage and repelled from the negative. The electrons in the P material near the junction pass quickly from hole to hole to the end of the P material where the battery positive is attached. They enter the battery.

As they leave the junction area, more electrons from the N material cross over the junction and continue toward the battery. The negative end of the battery helps by supplying more electrons into the N material. The additional electrons keep repelling the roaming electrons toward the junction and electrons pour over the junction into the P material. Then they pick their way through the P material from hole to hole and enter the battery, thus resulting in a steady current flow.

As long as the battery voltage is applied in the forward bias direction, a continuous current flows. The junction narrows down to practically nothing and the PN junction is a conductor. If too much current is caused to flow, it can overheat the PN junction and ruin the diode.

The diode in forward bias is not as good a conductor as a piece of copper. While copper has a resistance of a tiny fraction of an ohm at that small length, a diode has a forward resistance of a few ohms. This is a few hundred thousand times more resistance, even if it is a few ohms. Therefore, too much current flow creates a lot of heat and the diode breaks down.

To sum up, due to the variable barrier width characteristic, the PN junction exhibits excellent diode action; that is, it permits current flow when it is forward biased and halts current flow when it is reverse biased, just like a switch. The amount of bias must be within the rated characteristics of the diode. Too much bias in either direction can cause the diode to be destroyed. A PN junction is a piece of N material with an excess number electrons joined to a piece of P material with a deficiency of electrons or the presence of holes.

TESTING THE JUNCTION

The P material is always the anode of the diode. The N material is always the cathode of the diode. The diode schematic symbol (Fig. 1-11) is a line and a triangle. Try to get away from thinking of the triangle as an arrowhead, for it points in a confusing way. If you consider it as a triangle, then it's easy to remember it's the anode. With forward bias, current always flows from the cathode line to the anode triangle.

The most convenient way to test a PN junction or diode is with an ohmmeter. Assuming an ohmmeter is being used in which the battery is arranged so that the positive end is attached to the positive output terminal, conditions of forward and reverse bias can be applied across the diode (Fig. 1-12). With the positive probe attached to the P material and the negative probe attached to the N material, forward bias is applied. With forward bias the resistance of the diode, with a small battery voltage like 1.5 volts, can be a few ohms. The resistance will vary with the actual diode, but the typical

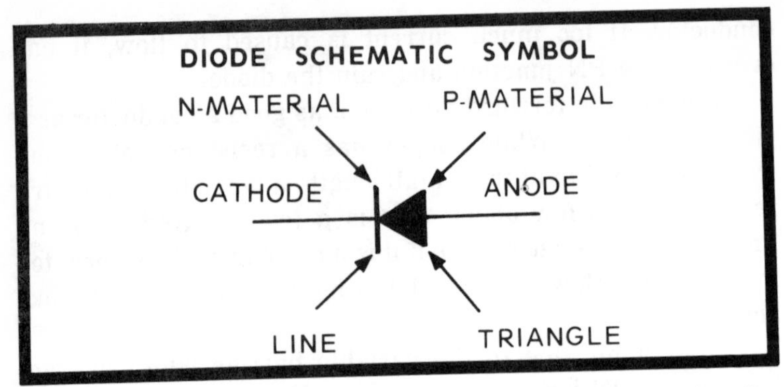

Fig. 1-11. Schematic symbol of a diode. The N material is represented by a line and the P material by a triangle.

silicon or germanium resistance is a few ohms, forward biased.

When the negative probe from the ohmmeter is attached to the P material (anode) and the positive probe is attached to the N material (cathode), the diode is reverse biased (Fig. 1-13). With reverse bias the diode has a resistance measured in the hundreds of thousands of ohms. These resistances become

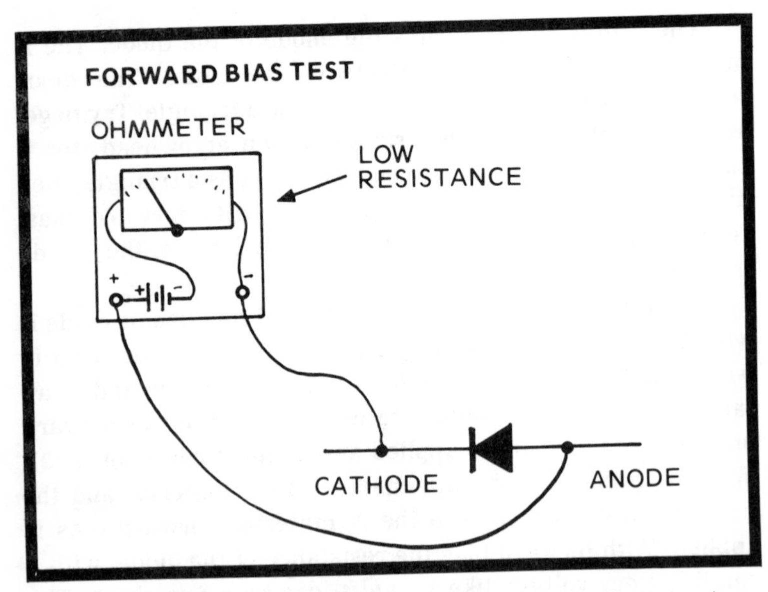

Fig. 1-12. When a diode is tested in a forward bias mode, the ohmmeter shows it is a very low resistance.

quite important when considering characteristics of transistors (see Chapter 2).

Why isn't the reverse bias higher? Why doesn't the resistance approach infinity and the amounts of current approach zero? Remember the minority carriers. The holes in the N material and excess electrons in the P material permit small amounts of current to flow in the reverse direction, lowering the reverse biased resistance. The following table should help fix these facts in your eyes.

TEST	BIAS	CURRENT FLOW	RESISTANCE
Attach ohmmeter positive to anode, minus to cathode	Forward	Heavy	Very low (few ohms)
Attach ohmmeter positive to cathode, minus to anode	Reverse	Minute	Very high (100K or more)

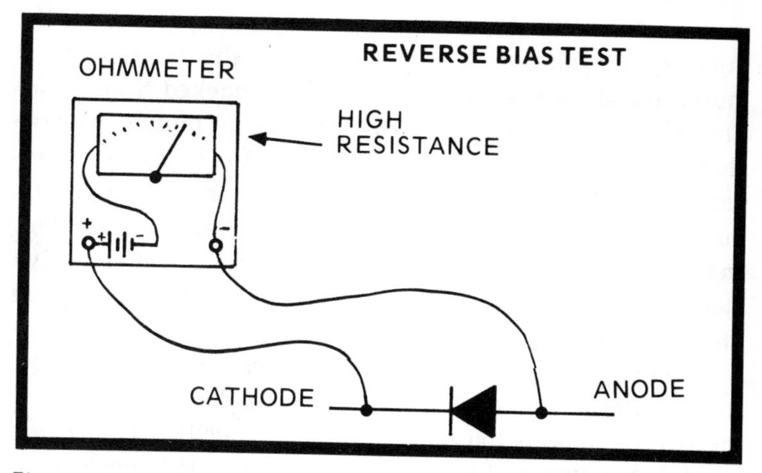

Fig. 1-13. When a diode is tested with a reverse bias, the ohmmeter shows a very high resistance.

CHAPTER 2

Diodes & Bipolar Transistors

In a typical silicon diode, a piece of N material, symbolized by a line and a piece of P material, symbolized by a triangle, are bonded together. When the diode is forward biased, a voltage of about 0.6 volts turns on the diode and current flows from the excess of electrons point in the N material, across the barrier through the P material to the deficiency of electrons point at the positive terminal (Fig. 2-1). Holes flow in the opposite direction, but in troubleshooting, the holes do not get very much consideration, if any. The negative charges, or the electrons, are the main consideration.

Silicon rectifiers can pass high current easily, even as much as 400 amperes, with voltage pressures up to 130 volts in a temperature environment equal to the boiling point of water. They are small, lightweight, shockproof and efficient.

When a diode is reverse biased, little current flows due to a forward-to-reverse resistance ratio of 100 to 1. The small amount of current that does flow is due to the presence of the minority carriers, the ten electrons in the hole-ridden P material and the few holes in the electron-packed N material. This minority current pathway permits a minute reverse electron flow in the order of only a few microamperes, compared to the full ampere amounts of current that pass during a forward bias condition. These favorable power parameters enable silicon diodes to be used as power rectifiers. Five hundred ma., one amp, 25 amps, 50 amps and so on are typical maximum average rectified currents.

Another important parameter is peak inverse voltage, which is the safe amount of reverse bias a diode can stand. If the reverse bias is too high (beyond a specified point), the crystal structure just breaks down and the diode becomes a very low resistance. Current flows unhampered in a reverse direction. The device loses all diode characteristics and becomes, in effect, a piece of wire. When the peak inverse voltage (PIV) is exceeded, a permanent short occurs in the diode and it is ruined (zener diodes are designed aound the PIV parameter. See the section on zeners in this chapter.

DETECTOR DIODES

While the PN junction exists in a detector diode, it has an entirely different physical mounting than in the power diode. The detector diode is typically made of germanium. The schematic symbol is the same as that used for a silicon power diode, but the actual device looks different.

A piece of germanium is joined to a catwhisker. (Yes, the old crystal detector is still in use in solid-state circuits. The old crystal radio was a solid-state circuit. Nothing has changed.) The point contact between germanium and catwhisker is a PN

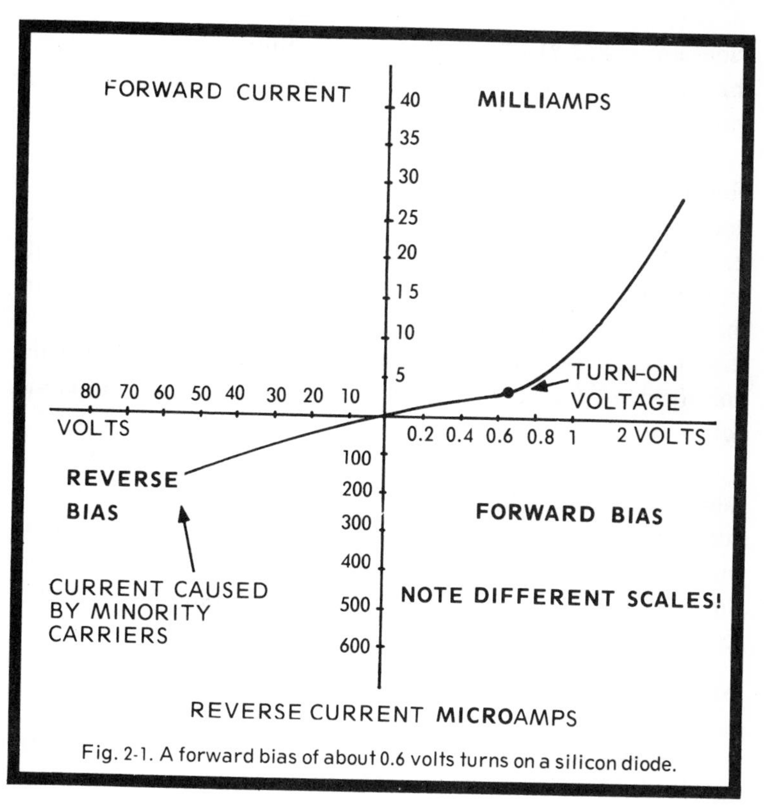

Fig. 2-1. A forward bias of about 0.6 volts turns on a silicon diode.

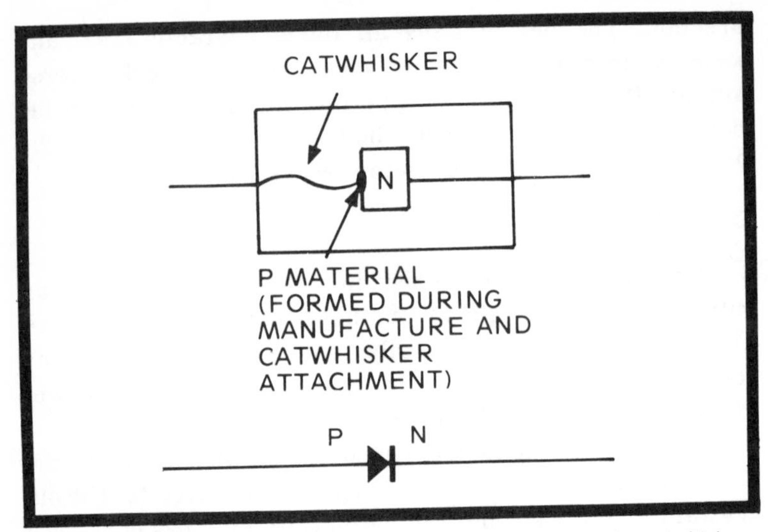

Fig. 2-2. A detector diode is actually a piece of N material and a catwhisker, which is nothing but a piece of wire.

junction (Fig. 2-2). The piece of germanium is N material and the catwhisker point contact constitutes, in effect, the P material.

A germanium diode turns on when the forward bias is closer to zero than the silicon. The germanium needs only about 0.2 volts in order to start conducting (Fig. 2-3). The germanium diode is used in applications where small amounts of current are involved, such as in detector circuits. The forward resistance is in the order of 100 or 200 ohms; therefore, current through the diode produces a small forward voltage drop. Currents in the order of zero to 50 milliamps are common.

A small amount of current flows in a reverse direction in a germanium diode, too, due to minority carriers. The reverse current flow increases almost in a linear fashion according to the applied reverse voltage. This is unlike the silicon where there is little reverse current until the breakdown voltage is exceeded.

DIODE CAPACITANCE

A diode junction is an insulator of sorts and, therefore, constitutes the dielectric of a capacitor (Fig. 2-4). The width of

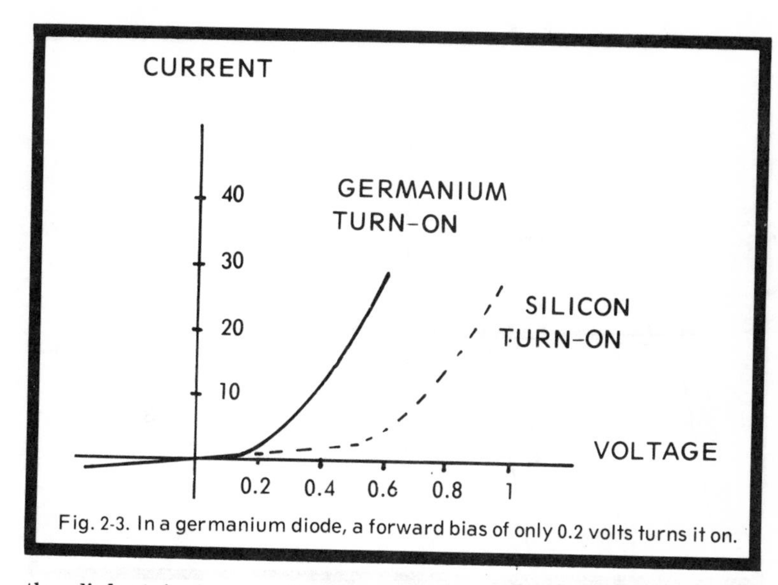

Fig. 2-3. In a germanium diode, a forward bias of only 0.2 volts turns it on.

the dielectric is practically nothing, and in a capacitor the further the dielectric keeps the plates apart, the less the amount of capacitance. The narrower the dielectric, the greater the capacitance. Therefore, since a diode junction is so narrow after the bond is made, the diode has relatively large amounts of capacitance.

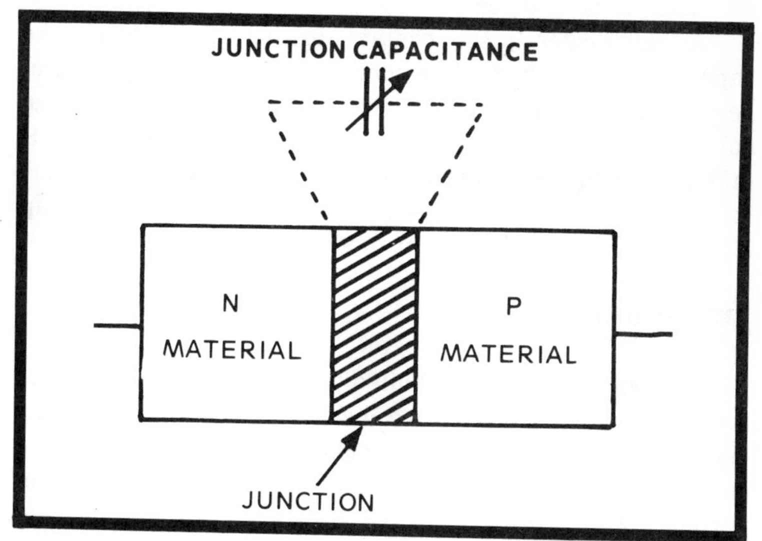

Fig. 2-4. The junction of a diode acts as a dielectric and gives the diode capacitance.

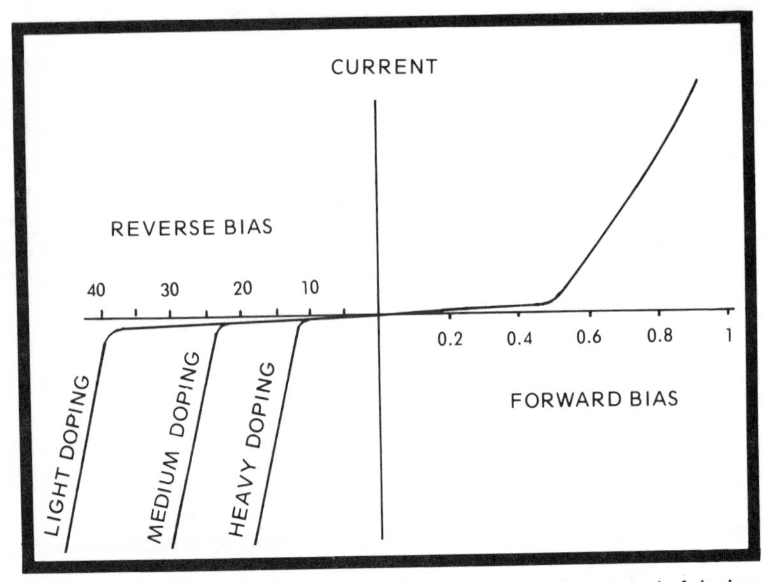

Fig. 2-5. A zener diode's voltage rating is determined by the amount of doping it receives.

The capacitance of a diode places a limit on the frequency at which it can work. In a silicon diode the capacitance is large. But since it operates as a power rectifier at a typical 60 Hz, there is no problem.

A catwhisker point contact drastically reduces the capacitance of a diode. The junction area is so tiny it compensates for the narrowness of the dielectric. A capacitance of only a few picofarad or less is present. Actually, the P material region is a microscopic dot on the N piece of germanium. Diode capacitance is put to good use in varicap diodes, covered further on in this chapter.

ZENER DIODES

The zener is a specially built diode that takes advantage of the reverse voltage breakdown characteristics. It has no other use, does not rectify, or anything else another diode is capable of doing. Zeners are made of silicon and are available in various power ratings. However, they are referred to as a 9-volt zener, 12-volt zener, 50-volt zener, 200-volt zener and so forth (Fig. 2-5). The voltage designates the point in the zener characteristic curve where the reverse bias causes a sudden change from no current flow to maximum current flow (Fig. 2-6). Zeners are designed not to break or short when the reverse current starts to flow. The sudden change in the graph curve looks like a "knee." Therefore, that voltage is called the zener knee voltage.

Once a zener begins conducting the voltage across it becomes fixed at the knee point. No matter how much voltage you apply to that circuit, as long as it is more negative than the knee voltage, the knee voltage drop remains the same. More current can be pumped through and larger inputs of voltage will cause current changes, but the voltage won't waver.

The zener diode is placed into the circuit with its N material or cathode attached to the positive voltage point or B+. The P material or anode is attached to the more negative spot. Therefore, the diode is reverse biased. If it's a 9-volt zener, it keeps the B+ at 9 volts at all times. If the voltage tends to rise above 9 volts, the knee voltage is reached and current flows from the more negative anode to the more positive cathode and thereby regulates the voltage at 9 volts.

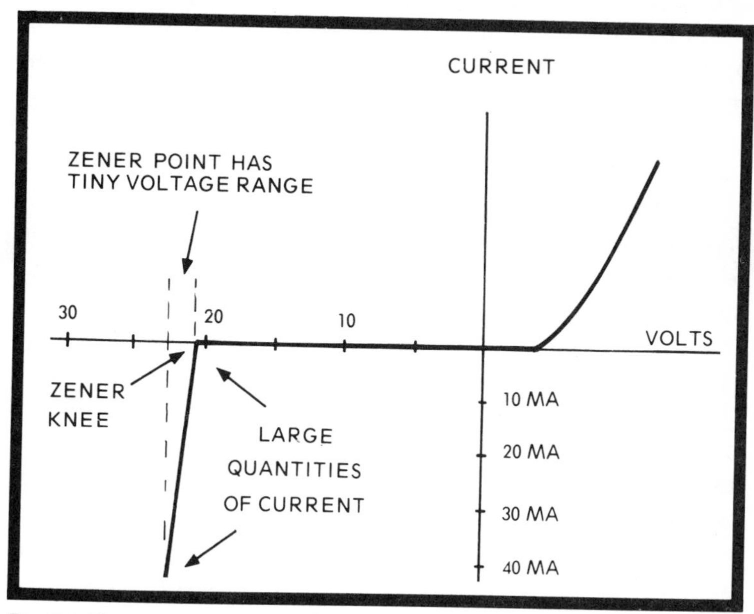

Fig. 2-6. The zener diode characteristic shows a "knee" at the reverse break-down point.

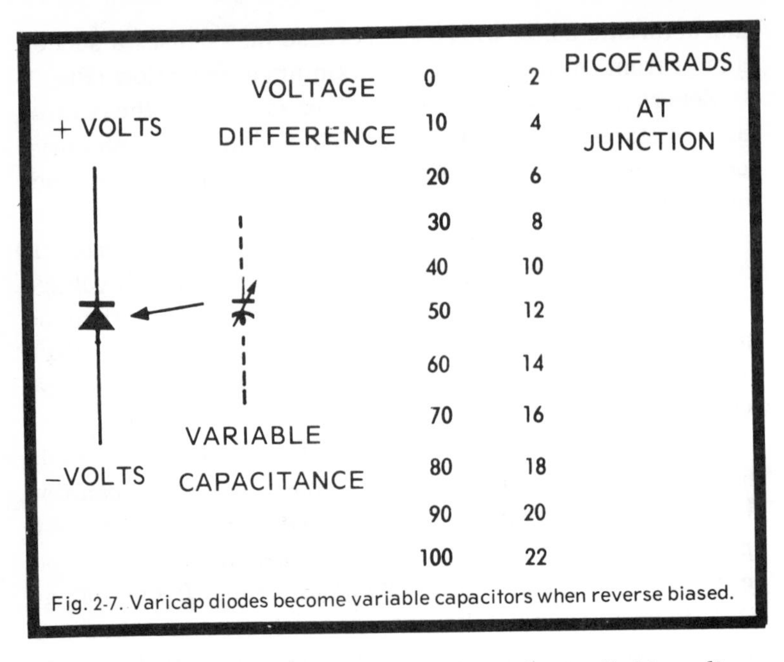

Fig. 2-7. Varicap diodes become variable capacitors when reverse biased.

Zener diodes can be connected in series to divide voltage and regulate it at the same time. Power ratings go up to 100 watts and more. Adding the diodes in series improves the power handling capability and all voltage levels can be required.

VARICAP DIODES

As mentioned earlier in this chapter, there is a definite amount of capacitance across a diode junction, and if the DC reverse bias is varied, the width of the barrier is varied. Since the barrier is, in effect, a dielectric, a diode can be made to be a variable capacitance by changing the DC reverse bias (Fig. 2-7).

A varicap is exactly like any silicon diode. It can be used in normal diode activity. It becomes a varicap only when it is reverse biased. Then it functions an actual capacitor in the circuit.

Since the capacitance varies in a linear fashion according to the amount of reverse bias, the varicap is useful in feedback circuits to control such things as the frequency of a multivibrator and the tuning of LC circuits. Other applications include remote control of tuned circuits, sweep tuning and FM modulation.

The Q of a varicap must be high. If the Q of a varicap in an oscillator circuit is low, the output will drop drastically. The Q is determined by the amount of leakage current through the diode when it is reverse biased. If the leakage is excessive, the Q will be low. The Q is directly related to the amount of leakage. The less the leakage the higher the Q.

To sum up, diodes are simply pieces of N material and P material bonded together. They are made of silicon for power applications and germanium for detector type uses. The silicon diode exhibits the knee voltage characteristic at reverse bias, it also functions as a variable capacitance when reverse biased. The germanium diode uses a stiff piece of wire (a catwhisker) attached to a piece of N material, thus making a low-capacitance diode.

BIPOLAR TRANSISTORS

A diode is a simple PN junction, as we've learned. To make a form of transistor, all you have to do is bond another piece of either N or P material to the PN junction. If you add a piece of N, you have an NPN transistor. Should you add a piece of P material, a PNP transistor is formed. Of course, NPN and PNP transistors are not produced exactly in that manner, but for troubleshooting the idea is valid. The word, bipolar, refers the two end pieces of material, which are identical. For a PNP type there are two P pieces, and two N pieces are used in an NPN type. From here on, we will call the bipolar a transistor. This will differentiate it from the field-effect transistor or what we call the FET.

The center piece of material, which is different than the material in the ends makes both NPN and PNP transistors equivalent two diodes, back to back. This also means that the two sections each have all the characteristics of two diodes including a zener knee voltage and the variable junction capacitance for each.

The zener knee voltage is avoided, because if it is reached the transistor junction ruptures and ends the life of the unit. The varicap action is considered carefully in a lot of ap-

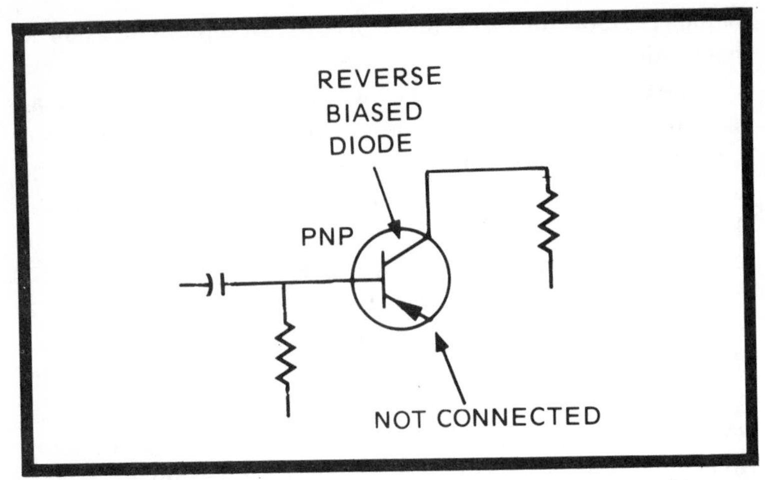

Fig. 2-8. A transistor can be used as a varicap by omitting the emitter connection.

plications, especially those in tuning circuits. On occasion, a transistor is used as a varicap by leaving the emitter connection loose and attaching only the base and collector (Fig. 2-8). The same type of application is used in some detector circuits in which the transistor functions as a diode.

The N and P materials have leads attached and are known as the emitter, base and collector, abbreviated E, B and C. A rough analogy used in the business compares the emitter to the cathode in a vacuum tube, the base to the control grid and the collector to the plate (Fig. 2-9). This indicates that electrons originate at the emitter, pass through the influence of the base and end up at the collector. However, the transistor is a current device, while a vacuum tube is a voltage device. No troubleshooter in a busy shop takes current readings. Everything is geared to correspond with voltage readings.

Also, on some transistors the collector voltage is negative with respect to the base and the emitter. How can electrons flow from emitter to collector in those cases? Well, they can't but holes can. Yet it's a better technique to envision electron flow rather than hole movement during troubleshooting. Why complicate the already tricky task.

An NPN transistor is the closest to the E, B, C-K, G, P analogy. Let's clarify the situation by examining the electron flow. An NPN transistor has an NP junction from emitter to base. The base is biased with a positive voltage and the emitter can be attached to ground. As in any good diode, electron flow goes from an area of excess of electrons to an area lacking electrons at the positive voltage on the base. Current flow takes place from the emitter to base (Fig. 2-10). That is what is meant when we say the transistor is a current device. The amount of bias on the EB junction determines the amount of EB electron flow. If a modulating signal arrives at the base and varies the bias, the amount of current will vary in accordance to the modulation. In other words, the modulation is varying the EB current flow (Fig. 2-11).

In a vacuum tube there is no current flow from the cathode (K) to the control grid (G1). The bias voltage on the control grid is negative in relation to the cathode. If a modulating signal arrives at the control grid, it varies the bias voltage just like it does on the transistor base. The varying voltage has a slight influence on the cathode ray passing through the control grid spaces; therefore, the G1 voltage varies the cathode-to-plate electron flow (Fig. 2-12). That is why the vacuum tube is called a voltage device.

In an NPN transistor, the collector is similar to the plate in a tube. In addition to a small current that flows from E to B, a heavy current flows from chassis ground (E), with its excess of electrons, to the positive charged C with its lack of electrons (Fig. 2-13).

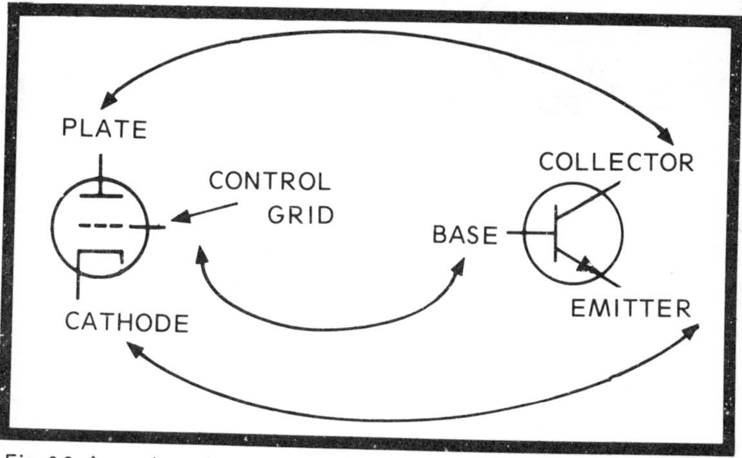

Fig. 2-9. A rough analogy compares a tube's cathode, control grid and plate with the emitter, base and collector (respectively) in a transistor.

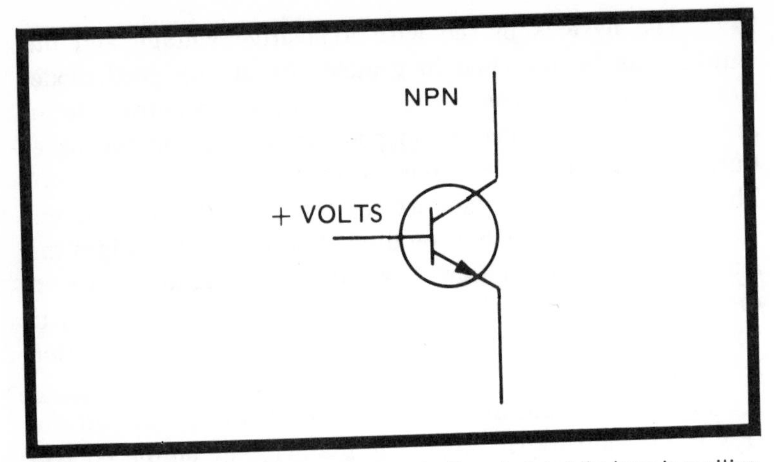

Fig. 2-10. An NPN is forward biased when the P material at the base is positive to the N material of the emitter.

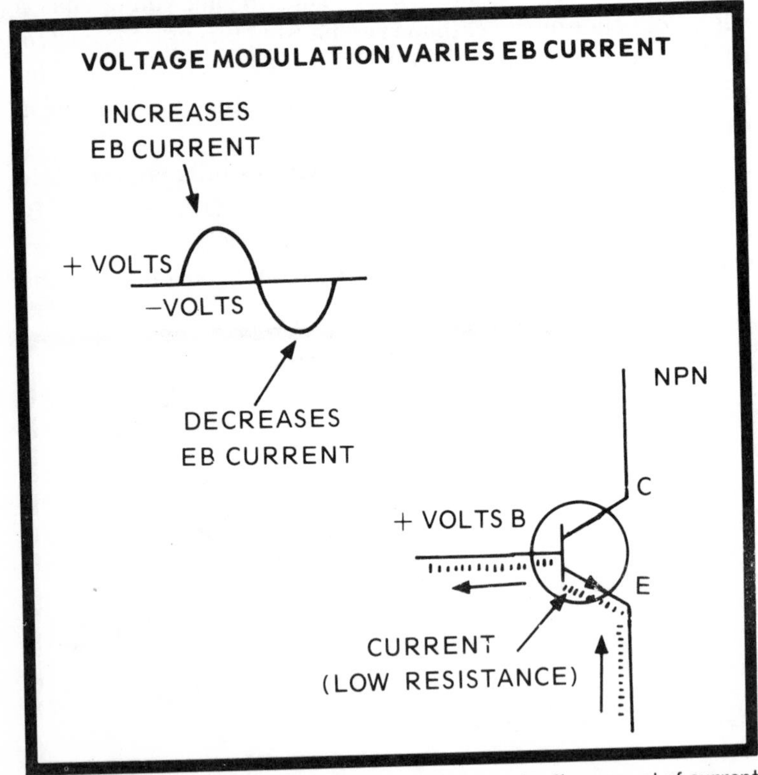

Fig. 2-11. A signal voltage applied to a transistor varies the amount of current flowing between emitter and base.

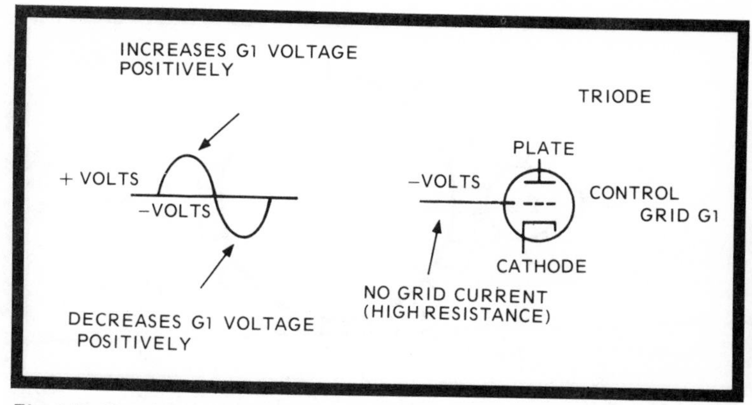

Fig. 2-12. Signal voltage applied to a tube varies the amount of voltage present at the control grid.

The EB current influences the EC current. Any modulation at B is also present at C. It's as if EB is a small gear that turns EC (Fig. 2-14). Any movement of the EB gear is amplified many times in the EC gear. This is due to the forward and reverse bias on the two back-to-back diodes. Across the EB diode the bias is forward, which means very little resistance. Practically no voltage drop takes place. A tiny bias gets the EB current moving. In germanium transistors, like any germanium diode, 0.2 volts bias starts the current flowing. In silicon transistors, 0.6 volts is needed. At first glance the troubleshooter sees almost identical E and B

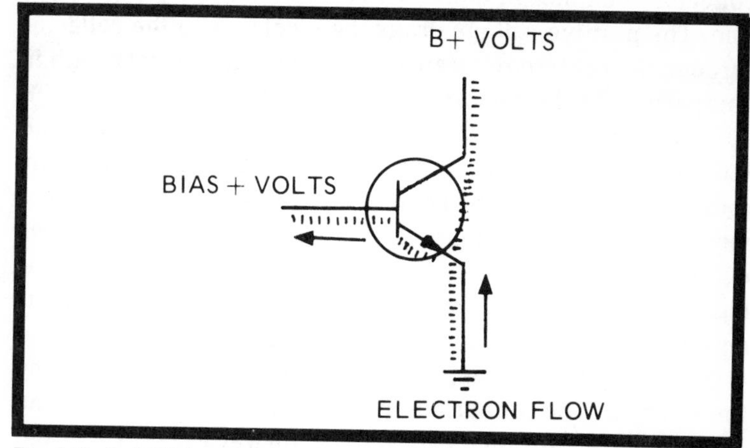

Fig. 2-13. An NPN transistor is like a tube in that electrons flow from chassis ground to the positive collector.

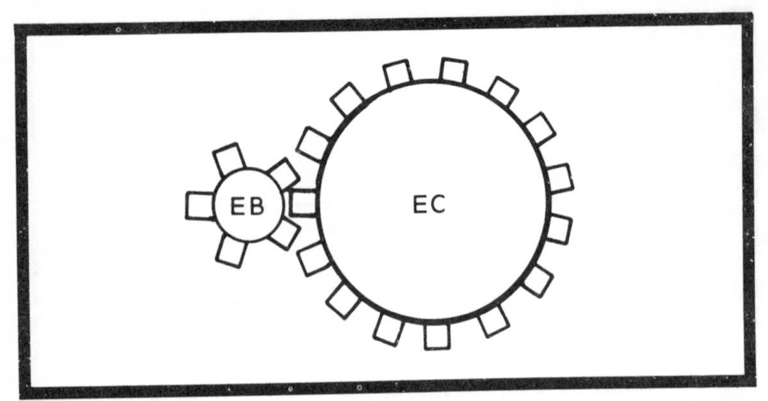

Fig. 2-14. Transistor amplification may be compared to two gears. The EB junction is represented by a small gear turning the large gear, represented by the EC junction.

voltages. A closer look reveals 0.2 or 0.6, as the case may be, on a forward biased transistor. Lots of circuits have the EB junction reverse biased. That way, no current flows until a signal arrives at the base to forward bias the junction. A transistor cannot have a high positive bias on the EB junction; otherwise, the danger of too much current exists. It's similar to a vacuum tube that has too much positive voltage on the G1 and runs too hot.

The BC junction in an NPN is reverse biased. C has a B positive voltage on it. Electrons are attracted strongly to the collector even though there is a reverse bias on the BC junction. The positive voltage drags the electrons to the collector through the reverse resistance. The resistance is very high in comparison to the forward biased EB junction.

The currents that flow from E to B and E to C are about the same. The resistance from E to B, forward biased, is in the order of a hundred or two ohms. The resistance from E to C, including the reverse biased B to C, is on the order of hundreds of thousands of ohms (Fig. 2-15). Therefore, the voltage drop is high and the amplification gains are great. This is why there is amplification in a transistor. The small modulation that arrives via the base appears in the collector circuit many times larger. The voltage amplification is a direct result of the difference in resistance between the forward biased EB and the reverse biased BC junctions.

While the voltage applied to an NPN can be compared to a tube circuit, that is, a positive voltage on the collector and B+ on the plate, a PNP transistor operates with a negative voltage on its collector (with respect to the emitter). The previous analogy is accurate, however, if you consider hole movement in a PNP instead of electron flow. The holes move from the emitter to the collector and the emitter to the base (Fig. 2-16). However, in my opinion it is easier to relegate the thinking about hole movement to deeper studies and forget about them in troubleshooting. Use electron flow in PNP as well as NPN types.

In a PNP circuit, electrons originate at the negatively charged collector. The electrons flow from the collector to the base and also in heavier numbers to the emitter. This is exactly the opposite to the way a cathode ray flows in a tube. It's like thinking of electrons flowing from the plate to the cathode. While this does not occur in a vacuum tube, it does occur in a PNP transistor.

In the collector load resistor the electrons move the other way (Fig. 2-17), but there is just as much of a voltage drop

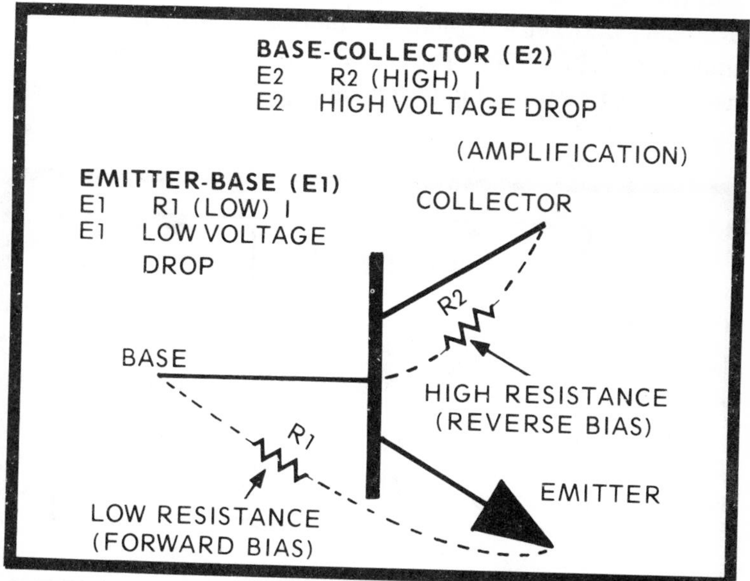

Fig. 2-15. The currents flowing from C to B and E to C are the same. The difference in resistance causes the separate voltage drops and the large voltage amplification.

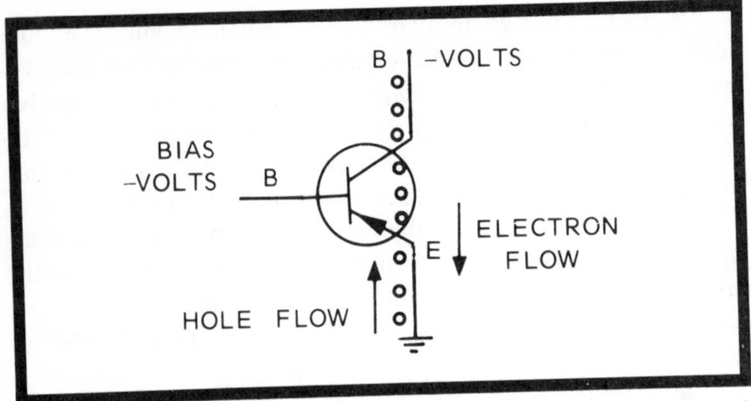

Fig. 2-16. In a PNP, holes move from E to C. Electrons, though, travel from C to E.

across the load resistor. The signal can just as easily be processed by passing electrons from a negative point in the power supply, through a load resistor, into a collector and out an emitter to the chassis. The base can just as easily modulate the C to E electron flow by varying the C to B electron flow.

The zener point of each individual diode in the transistor is still the high reverse bias voltage; the polarities are simply switched. Between each junction there is a variable dielectric that changes its width with the bias. The junction capacitances are a strong factor in higher frequency applications and must be considered.

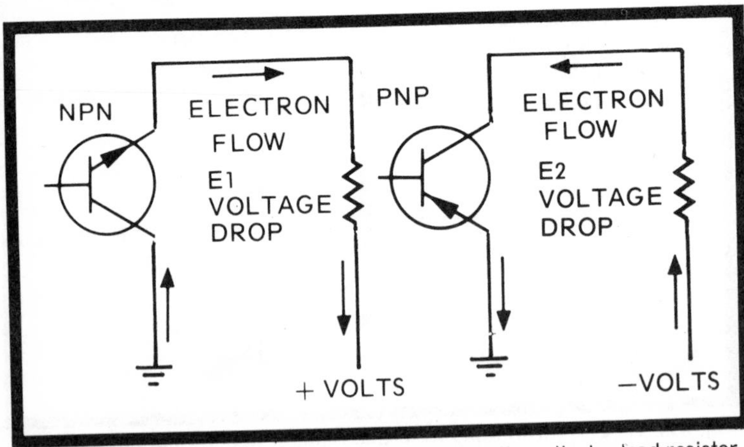

Fig. 2-17. The same voltage drop takes place across the collector load resistor, even though the electrons are going the other way.

In comparing a PNP transistor with an NPN type, all that has happened is the N and P materials have been switched. The EB junction must be forward biased. Since the emitter is attached to P material and the base is attached to N material, forward bias means a more positive voltage on the emitter (Fig. 2-18). The BC junction must be reverse biased in order to attain the high reverse resistance to current for a resultant voltage amplification. That means the collector must be negative with respect to the base.

Probably the most important part of the troubleshooting technique is a measurement of the voltages at E, B and C and the resultant current flow. Once you master this, the thinking part of a transistor circuit troubleshooting job becomes easier than it is with a comparable vacuum tube circuit. No pin numbers to look up, just E, B and C on either an NPN or PNP transistor. There are also some quick checks that can be made in a solid-state circuit. Let's go through the various tests that can be made on diodes and transistors.

TESTING DIODES

Whether a solid-state diode is doing power rectification, signal detection, voltage regulation, or acting as a variable capacitance, the general test is the same for all. The ratio of the forward current to the reverse leakage current has to be at least ten to one; the higher the ratio, the better the N and P materials are and the better the PN junction is.

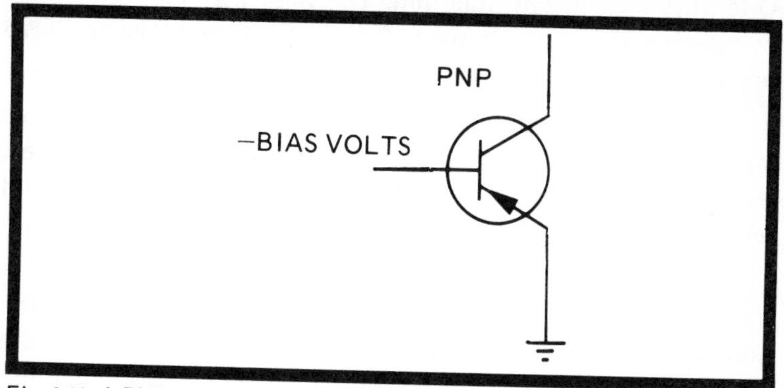

Fig. 2-18. A PNP is forward biased when the emitter is more positive than the base.

With an ohmmeter or a continuity tester with a flashlight bulb, a go no-go test can be made either with the diode in or out of circuit. The two poles are tested in first one direction and then the other. When the forward bias is applied, the ohmmeter reads a low resistance or the light goes on. When the reverse bias is applied the ohmmeter reads ten times higher or the light stays off.

If the ohmmeter reads the same in both directions, the diode is shorted if the resistance is low or open if it's high. If the continuity tester light stays on, the diode is shorted, or open if the light stays off. If the diode appears shorted while in a circuit, double check that there aren't any components in shunt with the diode. A coil in shunt or a shorted capacitor or resistor in shunt could produce a false short indication.

When testing a zener diode, if it should read shorted, double check that the battery voltage from the ohmmeter or continuity tester does not exceed the zener knee voltage. If it does, the zener will read short even though it's good, since it passes reverse current if the voltage applied is greater than its knee voltage.

Special care should be given to varicap diodes concerning the reverse leakage current. The Q of the diode is dependent on the reverse leakage. If the leakage is excessive, the Q drops. Should the leakage be almost non-existant, the Q rises. A diode that ordinarily would have some leakage, yet still provide diode action for other purposes, may have lost enough Q to hurt the varicap function. The best way to test it is by comparing a few varicaps and noting a know good reverse leakage.

Typically, silicon type diodes show practically no reverse leakage on the ordinary shop ohmmeter. Germanium diodes, on the other hand, can tolerate a bit of reverse leakage and will show some resistance.

Diodes can be tested in a more accurate (though not a more reliable way) with a good transistor tester like the Sencore TF 151. It can be tested in circuit by observation of the actual diode rectification, or out of circuit by reading the reverse leakage current in microamperes.

With a transistor tester a diode can be tested in-circuit by placing the type switch at NPN and attaching the E lead to the cathode and the C lead to the anode. When the tester is in the NPN mode, a positive voltage is present on the C lead and a less positive voltage on the E lead. This places forward bias on the diode. A current will flow similar to the current that flows from E to C in a transistor, except that it is higher in actual microamperes since there is no valve action from a base in a diode. A shorted diode will allow a heavier current flow and an open diode will permit no current flow. If there is no current or an extremely large amount, try turning the BETA CAL knob. If the needle moves at all to the right, the current flow is normal and the diode is good. If the needle doesn't do anything, the diode is either open or shorted.

The out-of-circuit test is the sure one. Instead of the EC junction probes, use the CB junction probes. The leakage across the junction in the suspect diode can be read in microamperes with the same test used for CB leakage in a transistor, since the PN junction for the test, is exactly like a CB junction.

With the type switch is set on NPN, bias is provided on the PN junction—a positive voltage on C and less positive voltage on B. If the B lead is on the cathode and the C lead on the anode, forward bias is provided to the diode. The current scale should read a large forward current in the leakage mode.

With the type switch on PNP, the bias is reversed. Any leakage current can be read directly on the microampere scale. As with the ohmmeter, the ratio should be at least ten to one or better for a good diode. As mentioned before, a silicon diode will show practically zero leakage. A germanium could read up to 50 microamperes and still possibly be good.

TESTING TRANSISTORS

Since a transistor is thought of as two diodes back to back (NPNs with anodes tied and PNPs with cathodes tied), a good reliable go no-go test can be made with an ohmmeter or flashlight continuity tester. First, one NP or PN junction is tested and then the other, exactly as if they were two separate

TYPICAL BETA VALUES

TRANSISTOR TYPE	APPROXIMATE BETA VALUES
RF AMP	50
IF AMP	100
AUDIO AMP	200
AUDIO OUTPUT	75

Fig. 2-19. Typical beta values.

diodes. Then another ohmmeter test is made from the emitter to the collector. It is quite possible that a short can exist between E and C even though the two diode junctions are good. This type of short is quite common and deceiving.

Another more accurate though not any more reliable test is a check of the transistor characteristics. A transistor is tested for beta and Icbo. Beta is the current amplification factor. This is different than vacuum tubes which are rated in terms of Gm in micromhos. Beta is derived from current measurements and is a final ratio figure. DC beta is the ratio of collector current divided by the base current. DC beta is given the cryptic hfe terminology. AC beta is the ratio of **change** in the collector current divided by the **change** in the base current, while keeping the collector voltage constant. It is called hfe. Forget the two letterings except when you look them up in a transistor manual. The best word to remember is beta. If beta measures about what the manual calls for, it is usually satisfactory (Fig. 2-19).

Icbo simply refers to the leakage current (I) between the collector and base (CB) when the third element, the emitter, is open (O). The Icbo is measured in microamperes and is probably the most important parameter of the transistor. It is what you are testing for in the go no-go tests. Excessive Icbo makes a silicon rectifier dangerous in a power supply and a

TYPICAL MICROAMP LEAKAGE VALUES

TRANSISTOR TYPE	LEAKAGE
SILICON RECTIFIER	ZERO MICROAMPS
GERMANIUM RF-IF	UNDER 5 MICROAMPS
VIDEO, AUDIO GERMANIUMS	ABOUT 20 MICROAMPS
POWER GERMANIUMS	ABOUT 1000 MICROAMPS

Fig. 2-20. Typical transistor leakage currents.

TRANSISTOR DC VOLTAGE TESTS

TEST POINTS	RESULTS	CONCLUSION
BASE TO GROUND	TOO HIGH OR TOO LOW	DEFECTIVE COMPONENT IN BASE CIRCUIT
COLLECTOR AND BASE	COLLECTOR AT SUPPLY VOLTAGE- BASE NORMAL (NO CONDUCTION)	DEFECTIVE COMPONENT IN EMITTER CIRCUIT
EMITTER, BASE AND COLLECTOR	EMITTER AND COLLECTOR IDENTICAL, BASE NORMAL	COLLECTOR HAS LOST SUPPLY VOLTAGE AND HAS DEFFCTIVE COMPONENT
EMITTER, BASE AND COLLECTOR	COLLECTOR AT SUPPLY, EB BIAS NORMAL (NO CONDUCTION)	OPEN TRANSISTOR
1-COLLECTOR	LOWER THAN NORMAL	GO ON TO 2-
2-SHORT E TO B READ COLLECTOR AGAIN	C DOES NOT RISE TO SUPPLY VOLTAGE	TRANSISTOR IS DEFECTIVE, IS LEAKING FROM E TO C.

Fig. 2-21. Transistor DC voltage tests.

zener diode useless, since it impairs the knee voltage. A high Icbo in a varicap causes the Q to drop drastically and stops a germanium diode from detecting.

The Icbo test is identical to the out-of-circuit diode test, except that the emitter lead is disconnected internally when you set up for the test. Although three probes are attached to E, B and C, only B and C are active. E is open. Icbo must be tested with the transistor out of the circuit. The only way the Icbo leakage test would be valid in the circuit is if there is absolutely no resistance across the transistor. Nine times out of ten, some resistance is present and you'll be reading the resistance rather than the transistor leakage. The leakage is read directly in microamperes (Fig. 2-20).

For all intents and purposes, beta and Icbo are separate characteristics. Quite often one will be bad while the other one is satisfactory. If either one is bad, the transistor is deemed defective. Other design characteristics of transistors such as alpha, alpha cutoff, transit time, inter-capacitance and feedback are not useful during routine troubleshooting. Fig. 2-21 lists typical transistor DC voltage tests.

CHAPTER 3
FETs

Field-effect transistors are solid-state devices that are made of N and P pieces of material, and at first glance look like NPN or PNP bipolars. However, they are very different. In fact, an FET is like a vacuum tube in many ways (Fig. 3-1). A vacuum tube has a high input impedance, a transistor has a low input impedance, and an FET has a high input impedance. Vacuum tube gain is measured in micromhos (Gm), transistor gain is measured in beta, and an FET's gain is measured in micromhos (Gm).

A vacuum tube acts as a valve due to the voltage potential that varies the tube's electron flow. A transistor's valve action is produced by a small bias current that varies the electron flow. Valve action in an FET is produced by a voltage potential that varies the electron flow.

JFETs

From the above statements it is readily seen that an FET is more like a vacuum tube in characteristics than an ordinary transistor, even though it looks like a transistor. The analogy goes even further.

We have compared the E, B and C of a transistor with the cathode, control grid and plate of a triode tube. Yet there are great differences between the base and the control grid; namely, the base exerts current control while the control grid exerts voltage control over the electron flow in the device. In the FET, this difference is resolved. Both the tube's control grid and the FET's corresponding element both exert a voltage control over the electron flow.

This brings us to the FET's elements and their names. They are the source, gate and drain (Fig. 3-2). Abbreviated S,

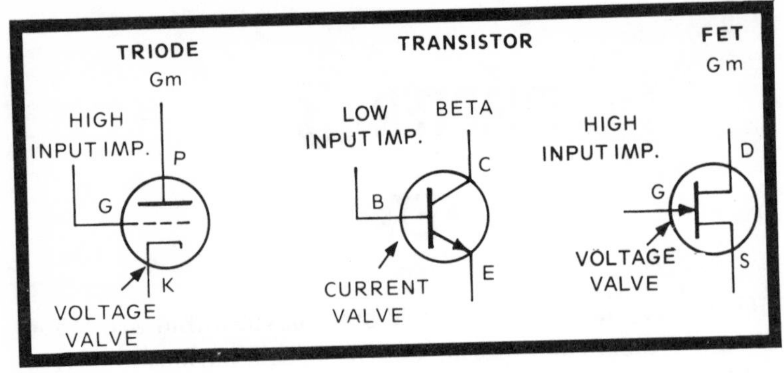

Fig. 3-1. An FET operates more like a vacuum tube than a bipolar transistor.

G and D, the source corresponds to the emitter of a transistor and the cathode of a tube. The gate is like the base and the control grid, and the drain is like the collector and the plate. However, at this point the FET becomes its own device. It is constructed in a unique way. A piece of P material can be the bottom of the device, called the substrate. A crevice is carved in the P material and in the crevice is placed a piece of N material, in which another crevice is carved out. In the N material crevice a small piece of P material is placed.

Consider the substrate as a rectangle (Fig. 3-3). The N material in its crevice stretches from one end of the substrate to the other. The piece of P material in the small crevice on top of the N material is simply a plug. Yet a "sandwich" has been made with a long piece of N material between two pieces of P.

The N material is called the channel. Therefore, such a device has an N channel. If the "sandwich" materials had been reversed with N as the substrate and plug, with a P

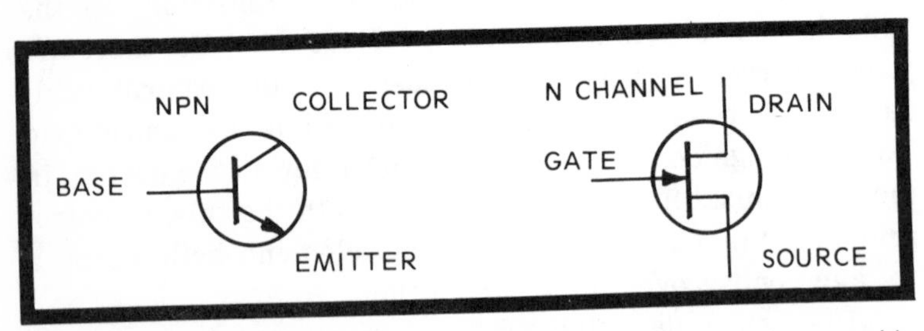

Fig. 3-2. An FET has a source, (S), gate (G) and drain (D) that correspond to the emitter, base and collector in a bipolar.

material in the channel crevice, the device would have a P channel.

The source and drain connections are both made to the channel, but at opposite sides of the gate. The gate connection is made to the plug. On occasion, a second gate can be produced by making an attachment to the substrate. There are PN junctions between the plug and the channel and between the channel and the substrate (Fig. 3-4).

This device is called a junction FET or more commonly an FET. An N-channel JFET works like this: A positive voltage (B+) is applied to the drain. The source gets a less positive voltage (Fig. 3-5). Current immediately begins to flow from the source, which has an excess of electrons, to the drain which lacks electrons. There is no barrier for the electrons to hurdle. They flow unrestricted, except for the actual resistance of the piece of N material which comprises the channel.

A negative bias voltage, exactly like that applied to a tube's control grid, is applied to the gate or gates if there are two. While there is no junction between the source and drain channel, there are PN junctions between the gates and the channel. With a negative voltage on the gate attached to the P

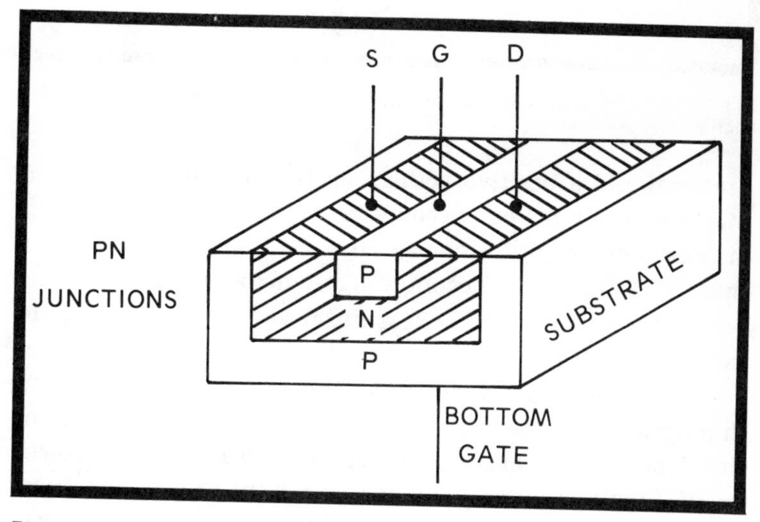

Fig. 3-3. An FET is constructed something like a transistor, but the connections are at different places.

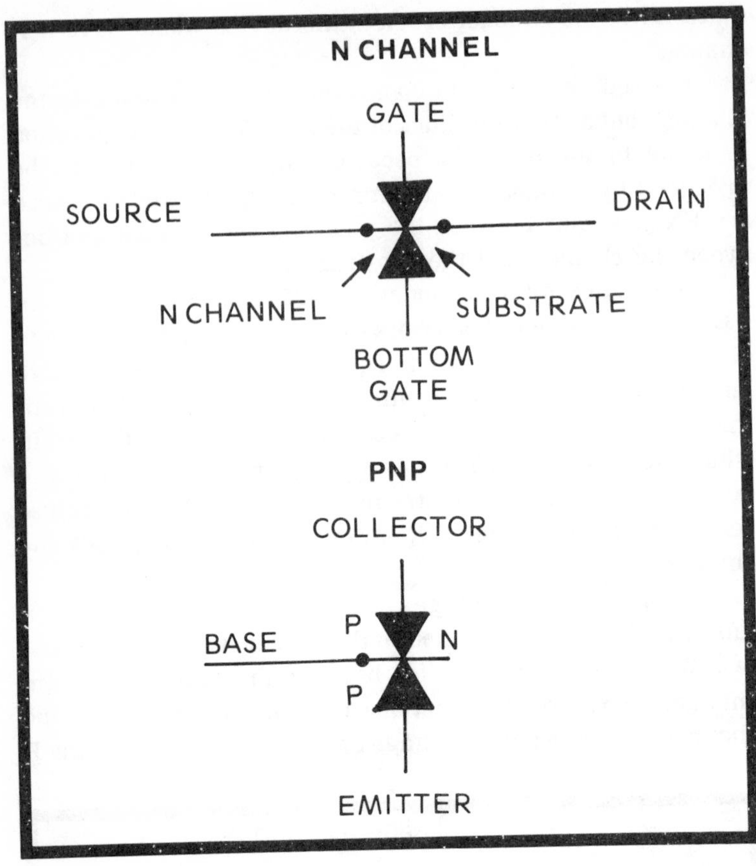

Fig. 3-4. An FET is also like two diodes connected back to back, but the connections are different.

material and a positive voltage on the channel, which is N material, the junctions become reverse biased. According to the size of the negative charge, the width of the junction will change. If the charge becomes highly negative, the PN junction will get quite wide. A less negative voltage on the gate narrows the junction.

The wider the PN junctions become, the less area the channel electrons have to flow in. This restricts the number of electrons that are going from source to drain. The narrower the PN junctions become, the more area the channel electrons have to flow in. This increases the number of electrons that are going from source to drain. A varying voltage on the gate, in turn, varies the electron flow in the channel, just as a control grid in a tube varies the cathode ray.

If the gate voltage becomes negative enough, it pinches off all the current flow. This is exactly like tube cutoff. The input resistance of a JFET is always high since there is always a reverse bias on the junctions.

In a P channel, the source and drain is attached to either end of the P channel and the gates are attached to the pieces of N material, the plug and the substrate. The drain is given a negative voltage and the source gets a less negative voltage (Fig. 3-6). Holes are attracted from the source to the drain. Electrons, however, move opposite to the holes and go from the more negative drain to the less negative source.

In order to reverse bias the PN junctions, a positive voltage has to be applied to the gates. Then, as the gates are made more positive, the PN junctions increase in width. As the gates become less positive the PN junctions become narrower. As the junctions become wider the resistance to the electron flow increases and the number of electrons that pass in the channel are reduced. Also, as the junctions become narrower, the channel widens and lets more current flow. At a very high positive gate voltage, the PN junction becomes so

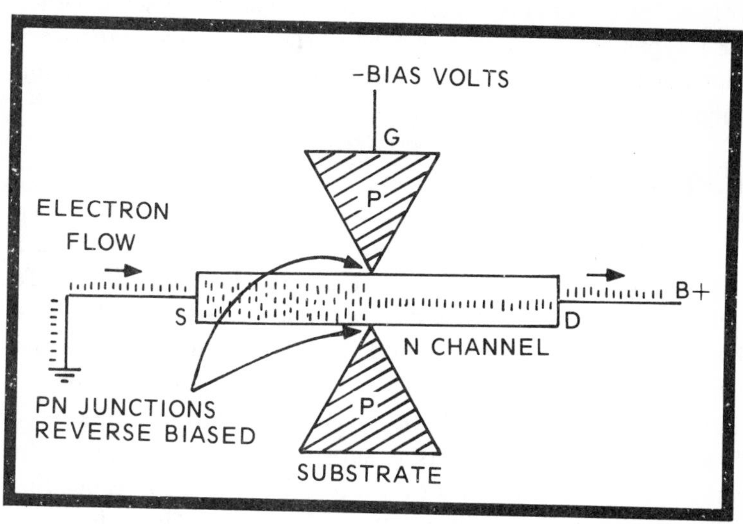

Fig. 3-5. With B+ on the drain and a negative bias on the gate, the electron flow can be modulated.

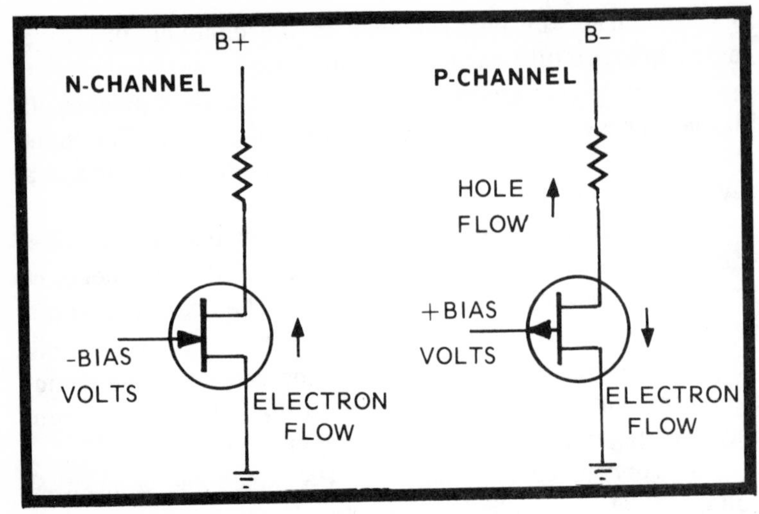

Fig. 3-6. The channel material used determines the polarity of the DC voltages.

wide that the channel is pinched off and no current flows. This is exactly like the action of an N channel, except everything is of the opposite polarity.

JFETs were the first type of FET to appear. The ones just described were of the depletion mode. As the gate voltage varies, the channel is depleted to a more or less degree. Other FET types are IGFETs and MOSFETs. They are transistors that not only operate in the depletion mode, but also in the enhancement mode.

IGFETs & MOSFETs

The IGFET stands for Insulated Gate FET. The MOSFET is a Metal-Oxide-Semiconductor FET. IGFETs and MOSFETs are practically the same devices and the names quite often are interchangeable. IGFETs and MOSFETs have the same connection names as JFETs, the source, gate and drain. IGFETs do not have three pieces of material making a "sandwich." There is only a substrate and a channel. The substrate can be either N or P material with the corresponding opposite material in the channel (Fig. 3-7). There is a PN junction between the substrate and the channel. The gate is made of a piece of ordinary conductor attached to the channel but insulated from the channel by a piece of dielectric material. The dielectric is usually the oxide of silicon (SIO2), called silicon dioxide. (Actually, it is glass.) It gives the input resistance of the IGFET a very high value, which is in the millions of megohms. This is a distinct advantage over the FET input which is in the hundreds of megohms.

While the JFET input resistance is a function of the bias, since it is a result of the width of the PN junction, the piece of glass in the IGFET has a steady high input resistance with no chance of back leakage unless the insulator itself breaks down. Otherwise, the IGFET is identical to the JFET. It has a source, gate and drain and it is an N- or P-channel type.

N-channel devices are negatively biased and the P channel is positively biased. This is because the N channel has a positive charge to pass electrons from source to drain, while the P channel has a negative charge that passes electrons from drain to source. (Holes pass from source to drain.) These are all depletion mode devices. A channel exists and the bias depletes the channel by widening the PN gate junction in the JFET or widening the insulated gate junction in the IGFET.

An enhancement mode device has no channel. The piece of material is lying in the substrate crevice, but it is designed to pass no electrons during a normal gate bias situation. That is when the enhancement mode is biased at zero or reverse bias. In an enhancement mode device, current flows only when a

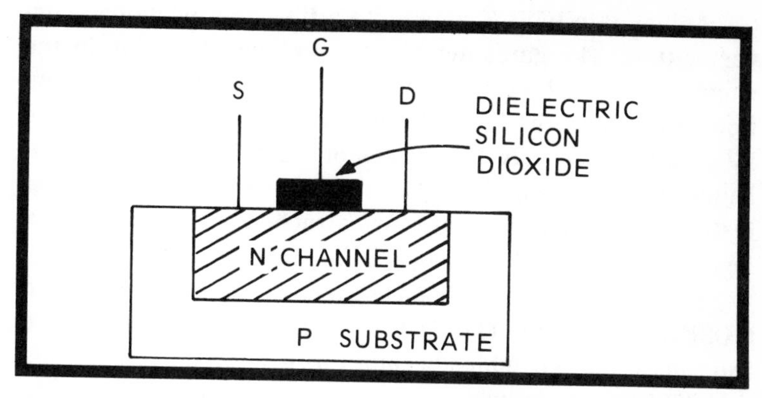

Fig. 3-7. In an IGFET, the gate is insulated from the channel by a piece of dielectric.

forward bias is applied. The forward bias opens the channel or forms the channel. It enhances the channel to a more or less degree, in contrast to the other type that depletes the channel to a more or less degree.

Enhancement type FETs are always IGFETs. A JFET would start acting like an ordinary transistor with emitter-to-base current flow. In an IGFET, due to the dielectric between the gate and channel, there cannot be any current flow from channel to gate. Therefore, a forward bias can be applied to get a channel current moving satisfactorily.

To boil the definition down, in a depletion mode IGFET a current is flowing in the channel all the time, except when the reverse bias is high enough to completely deplete the channel. An enhancement mode IGFET has no current flowing in the channel until a forward bias is applied to create and enhance a channel so current can start flowing.

Typically, depletion mode FETs are used in amplifiers, oscillators, etc., just like a triode or pentode tube. Enhancement mode IGFETs are typically used in switching circuits, since they are normally off and switched on with a forward gate bias.

IGFETs have been thought of as single gate devices. The same type of device when given a second gate sometimes is referred to as a MOSFET. This, however, is no hard and fast rule. The names from a pure technical point of view are interchangeable. The gate dielectric for both is a metal oxide.

A dual-gate IGFET acts quite a bit like a pentagrid converter tube. The gates act like the two control grids in the vacuum tube. Either one can modulate the channel current flow or pinch off the flow with the proper bias. The dual-gate MOSFET is still only two pieces of material, an N and a P bonded together (Fig. 3-8). One is the substrate while the other is the channel. The two gates are attached in the channel but insulated from the channel by the piece of glass dielectric.

The channel can be tapped between the gates, forming two MOSFETs in series. The first MOSFET starts at the source and ends at the tap, which becomes its drain. The gate for this first MOSFET is between this source and drain. The second MOSFET starts at the tap, which becomes its source, in addition to being the drain for the first one. Then the second MOSFET ends at the end of the channel where the final drain is attached. The gate for the second MOSFET is between the center tap and the final drain. The dual-gate MOSFET can be used not only as a pentagrid converter, but also as two triodes in series, among other things. It is a valuable device.

When compared to vacuum tubes, FETs are exactly twice as complicated. This is because tubes can conduct only from cathode to plate. There are triodes, pentodes, etc., used in amplification, switching, etc., but the electron path is still from cathode to plate.

N-channel FETs are like vacuum tubes in that they conduct from source to drain. There are JFETs, IGFETs, with single gates and dual gates with depletion and enhancement modes, used in amplication, switching, etc. (Fig. 3-9). However, a new concept, not shared by vacuum tubes, is the P-channel FET. The electron path is from drain to source. The P channel becomes confusing to the technician well versed in vacuum tubes, because everything is backwards (Fig. 3-10). The electron path in the wrong direction needs gate bias of the opposite polarity and B values below ground in the negative direction. Understanding P-channel devices requires a little extra mental effort until they become familiar.

TESTING FETs

The FET is quite like a vacuum tube in operation. The parameters that are measured during testing, aside from

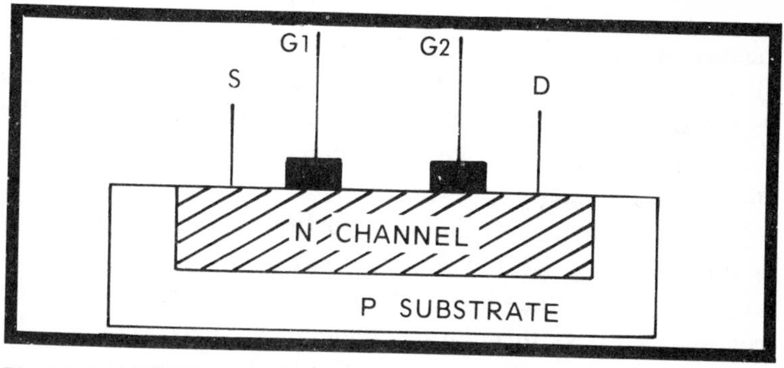

Fig. 3-8. A MOSFET often has two gates on top of the channel for dual control purposes.

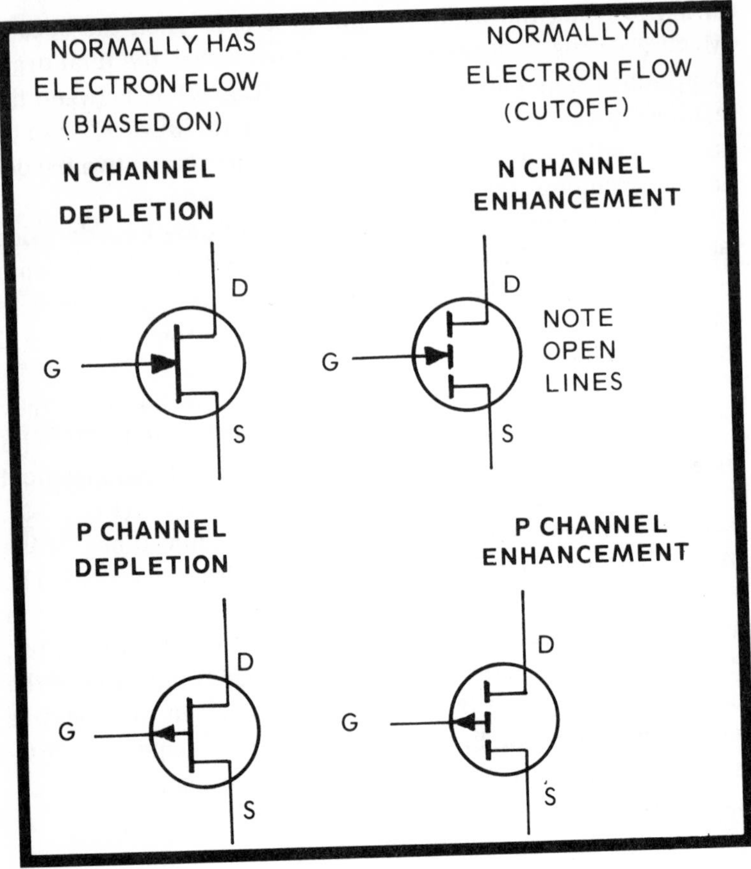

Fig. 3-9. MOSFETs are available in these four types. The depletion mode is useful as an amplifier, and the enhancement mode is used for switching purposes.

heater considerations which are non-existant in an FET, are like a vacuum tube. The gain of an FET is measured in the familiar transconductance (Gm) in micromhos, not beta like an ordinary transistor. This is because the FET is a voltage device with a high input impedance.

Another parameter that is important to the troubleshooter is the Igss (Fig. 3-11). This is the gate leakage, which is the same as grid leakage in a vacuum tube. It refers to how much current can pass through the dielectric. There shouldn't be any more than can pass through the dielectric of a glass capacitor, which is very little indeed. The term Igss stands for the amount of current (I) that can pass from the gate (G) to the source (S) with the other element the drain shorted to the source (S).

If the FET is a dual-gate type, both gates must be tested. The term Ig2ss is used, which refers to the amount of current (I) that can pass from gate two (G), to the source (S), with the third element (the drain) shorted to the source (S). The leakage is measured in microamperes.

A third parameter is Idss (Fig. 3-12). This is an important parameter in matching FETs. It is zero bias drain current. It is the amount of current (I) that can pass from the drain (D) to the source (S) or vice versa, with the third element (the gate) shorted to the source (S). With the gate shorted to the source,

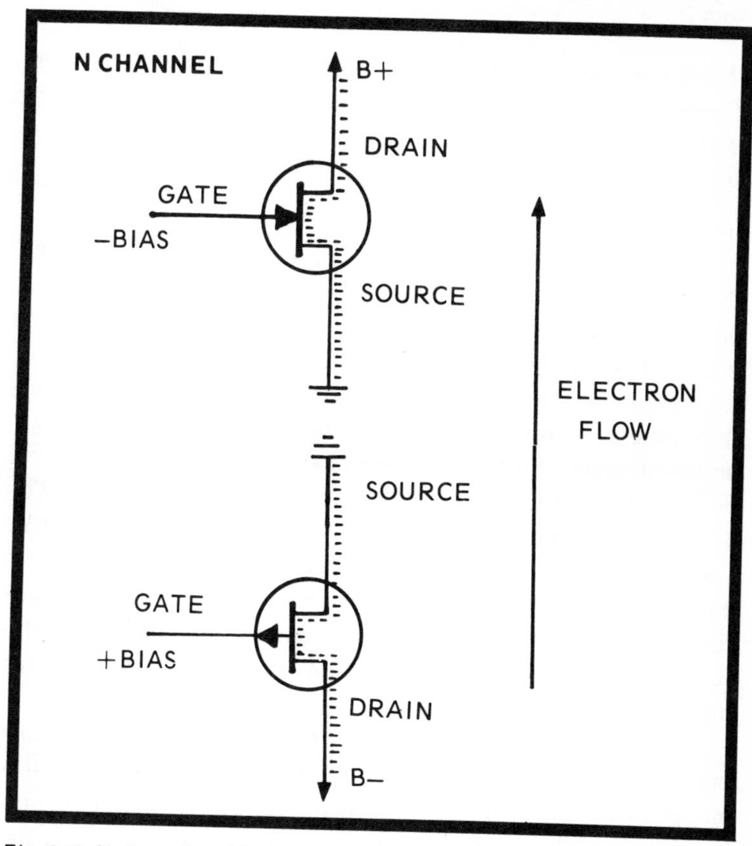

Fig. 3-10. N-channel and P-channel devices are mirror images of each other, above and below ground.

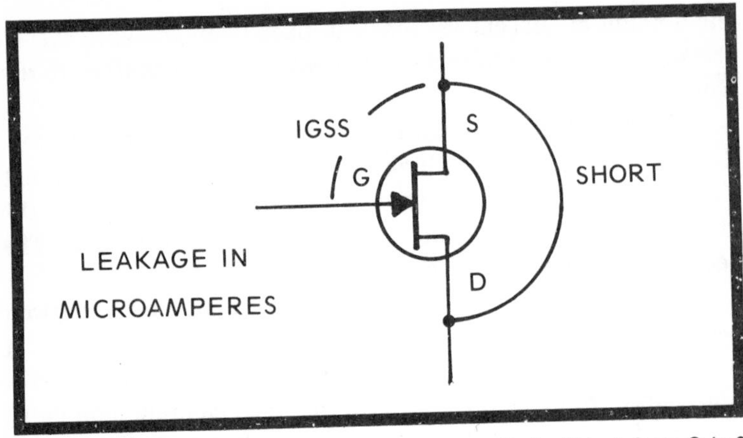

Fig. 3-11. The amount of current in microamperes that will leak from G to S with D shorted to S is Igss.

it cannot exert any influence over the channel current. Therefore, the bias is zero. A certain amount of current (in ma) will flow with zero bias. The current in milliamps is the zero bias drain current.

Handling

There are no safe go no-go tests for insulated gate FETs. The tiny piece of dielectric is subject to rupture at the slightest bit of electric charge. Even the static charge you build up as you walk around the room can destroy it.

When the IGFET is in its circuit, it is relatively safe, since it is properly grounded and attached. Out of circuit though, unless stringent precautions are observed, chances are it will be ruined in ordinary handling during routine shop work. The JFET, on the other hand, is just as rugged as any transistor or diode. These precautions are not necessary. On the IGFET in a replacement package, you'll notice there is a little shorting ring holding all the leads. **Do not** remove the short until the IGFET is in a circuit. The circuit could be the transistor tester input leads.

The insulation in the gate necessitates the unheard precaution. The problem is all centered around static charges and proper grounding. If you walk across a room on a dry day, build up a static charge from a deep rug and then touch the rabbit ears on a TV set, a spark will fly. Should the TV tuner have IGFETs directly in the antenna line, the static discharge could rupture the dielectric. That's how sensitive they are.

To properly unpackage and install an IGFET you must follow these instructions or chances are good you'll destroy the device. First of all, ground everything in sight. That means your hand, the soldering iron and the circuit where the IGFET is to be installed. Your hand can be grounded by attaching a jumper lead from your wrist watch or ring to the chassis; the soldering iron can be grounded by attaching a lead from the iron tip to the chassis. The chassis itself can be grounded by attaching a lead from the chassis to a conduit ground like the third hole of a wall socket. Of course, power must be removed from the chassis.

Do not remove the shipping short on the FET until it is completely soldered into the circuit. Don't use a soldering gun that clicks on and off. A soldering iron without the on-off action is essential. Keep the iron hot until the actual moment of soldering. Then pull the plug on the iron for the actual solder application. That way there is no current in the iron when it touches the IGFET. Use a heat sink and as little wattage as possible. A 30-watt iron is the maximum.

To sum up, a field-effect transistor is quite different than the ordinary transistor. The only similarity is the basic

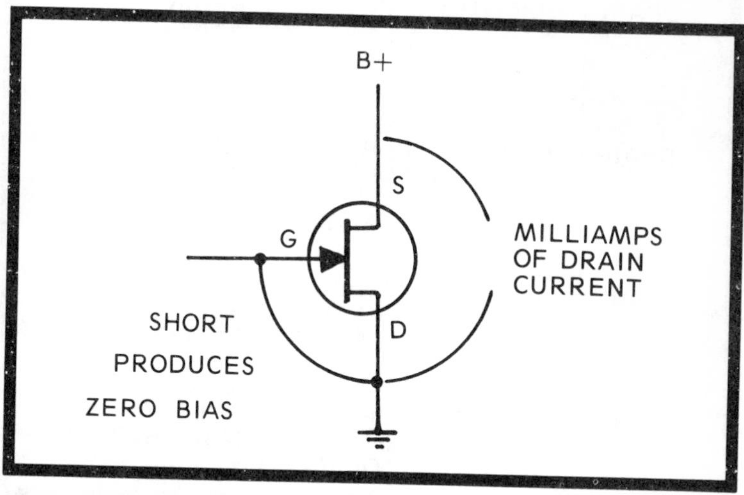

Fig. 3-12. The amount of current in milliamperes that will flow from S to D with G shorted to S is Idss.

building blocks of P and N material. The FET is more like a vacuum tube than a bipolar transistor. It combines the advantages of the vacuum tube and the transistor. It has a high input impedance like the vacuum tube and the instant on, filamentless advantage of the transistor. It is a voltage device with its gain parameter measured in Gm (conductance), like a tube, yet it offers the tiny power consumption of a transistor.

The JFET uses a conventional PN junction reverse biased to set up the proper channel bias. There are three pieces of material in the JFET, like a transistor, but the JFET does not sustain conduction across junctions like a transistor. The IGFET or MOSFET uses an insulated gate to set up the proper channel bias. As a result, the device can be biased in a reverse way like the JFET, but also forward biased to produce a new device in an enhancement mode. Handling of IFETs is critical out of circuit. Once in a circuit, they are as rugged as any other transistor. Fig. 3-13 is a DC voltage test chart for depletion mode FET operation.

TEST POINT	RESULTS	CONCLUSION
DRAIN	SLIGHTLY HIGH	DEFECT IN DRAIN LEG COMPONENTS
DRAIN	TOO HIGH, AT SUPPLY VOLTAGE	FET NOT CONDUCTING OPEN CIRCUIT IN SOURCE LEG, PINCH OFF BIAS TOO HIGH
DRAIN	SLIGHTLY LOW	DRAIN LEG COMPONENT, PINCH OFF BIAS TOO LOW, SOURCE LEG COMPONENT
DRAIN	MISSING	OPEN DRAIN LEG, OPEN GATE (NO BIAS), SHORTED SOURCE LEG

Fig. 3-13. Depletion FET DC voltage tests.

CHAPTER 4

ICs & SCRs

Two more common devices produced by the P and N building blocks are "chips" or integrated circuits and silicon controlled rectifiers.

INTEGRATED CIRCUITS

The integrated circuit is considered one of the wonders of the world. Entire stages are placed on pieces of material smaller than the nail on your little finger. One chip often contains a couple dozen transistors, with appropriate resistors and capacitors. However, the IC is not as all powerful as it would seem. There are major limitations that precludes its use as a replacement for all other electronic circuits.

When the IC can be used, it is valuable, especially in applications like a computer, where the same circuit has to be reproduced many, many times over. Instead of a computer filling up an office, as it would with tubes it can be placed on a desk. ICs are easily made containing diodes, transistors, resistors and capacitors, all hooked together in a useful circuit configuration (Fig. 4-1). The problems are few but major.

First of all, an IC has no pure inductance. So far there is no way to make an IC with a coil or transformer in it. Since this is one of the important parts of a circuit, universal use is immediately curtailed. There are ways to produce "artifical" inductance with amplifier circuits using reactance components. But, the lack of coils scratches a lot of applications. Some manufacturers simply build coils and attach them to the chip, but it's not the fully desired dream of the application (Fig. 4-2).

Secondly, resistors can be made on an IC, but their value and wattage are tiny. Unlike ordinary resistors produced

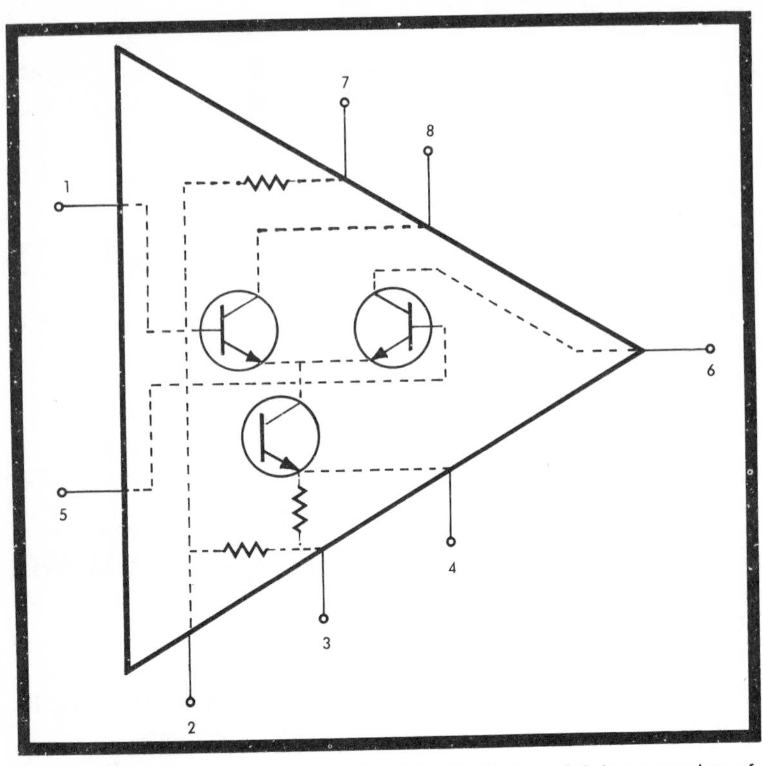

Fig. 4-1. An integrated circuit is a "black box" affair containing a number of circuits on a tiny chip of silicon.

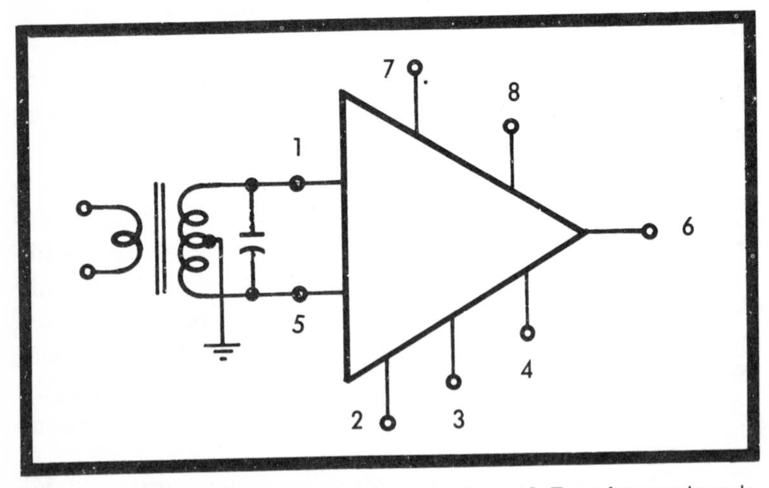

Fig. 4-2. No pure inductance can be wound inside an IC. Transformers have to be attached outside the unit.

easily in standard resistance and wattage ratings, the IC resistor depends on the amount of resistance in a particular block of material. High resistance requires a long, narrow area, while a low resistance utilizes a short, wide area (Fig. 4-3). Also, these areas are tiny indeed and the resistance and wattage must remain tiny.

Furthermore, while resistors on one chip can be made fairly correct in relation to each other, the resistance from chip to chip may vary even though they were made from the same design. Just a slightly different bonding of the building blocks changes the resistance.

IC capacitors are very small, also. The actual capacity is a function of the surface area of the dielectric. The amount of surface area available for the capacitor on the chip is microscopic. With a great deal of extra expense, the dielectric can be made thinner to produce more capacitance, but it's not too practical. The IC capacitor is said to be a function of

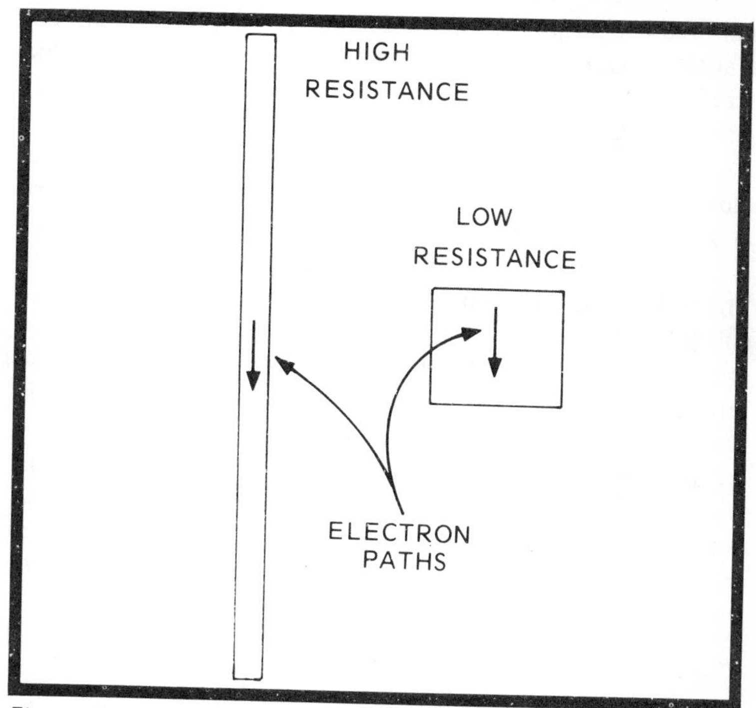

Fig. 4-3. Resistance is produced by narrowing, widening or lengthening the material electrons travel through.

surface area alone. Despite these restrictions, IC chips are valuable in many applications.

Chips are baked in ovens and typically start out with a piece of P material as the substrate, usually silicon. A mask is made like an ordinary printed circuit board mask, except that the mask is considerably smaller. Then, photochemical techniques are employed. To illustrate, let's quickly review a simple chip that contains an NPN transistor, a resistor and a capacitor. Of course, there are many more components on a real chip, but they are all made basically like this.

After masking the piece of P material and giving it a photochemical treatment, three pieces of N material are diffused into the P, with ordinary transistor-making techniques (Fig. 4-4). The size of the N diffusion is controlled by the oven temperature and baking time.

Next, another P area is diffused into the first two N areas just made. The third N area is let alone. Then, another N area is diffused into the first P area just made. This makes the number one area into an ordinary transistor with all the transistor characteristics. There are no limitations in the transistor part of the chip. The process is identical to normal transistor manufacturing.

With the P and N areas all formed, a layer of silicon dioxide is coated onto the chip. After that, metal contacts are attached to the appropriate spots. Now what do we have?

In the first diffusion area there is an ordinary transistor. The collector is attached to the bottom N material next to the substrate. The base is attached to the P material and the emitter is attached to the last piece of N material that was diffused. The emitter is attached to one end of the resistor; the resistor is formed over top of the P material above the second diffusion area. The other end of the resistor is attached to the capacitor. The actual capacitance is formed in the dielectric between the two capacitor connections. The chip is now a transistor with an emitter resistor and a capacitor attached to the emitter resistor.

TESTING ICs

Chips are encased in little packages resembling a transistor, but typically they have many leads; in fact, a dozen or more leads is not uncommon. ICs are usually represented on a schematic by a triangle pointed in the direction of signal flow. The leads come from the edges of the triangle with no schematic indication of what job the lead is doing or where it's coming from. The leads are usually numbered. Manufacturer manuals usually contain a "closeup" schematic representation of a typical IC. This shows the actual transistor, diode, resistor, capacitor configuration. It is a typical looking schematic. Also available sometimes are closeup photos, blown up many times, showing the way the substrate is metallized and where each component is located. For instance, the substrate pattern may be shown with all the test leads numbered; transistors are distinguished by Q designations, diodes by D, resistors as R and capacitors as C.

The chip can be tested by actually troubleshooting the configuration with appropriate signal injecting and signal testing techniques. In fact, this is the way a particular chip is tested after production in a factory. As each chip comes out of the oven it is hooked into a tester and a meter reads quality. One or more tests may be performed.

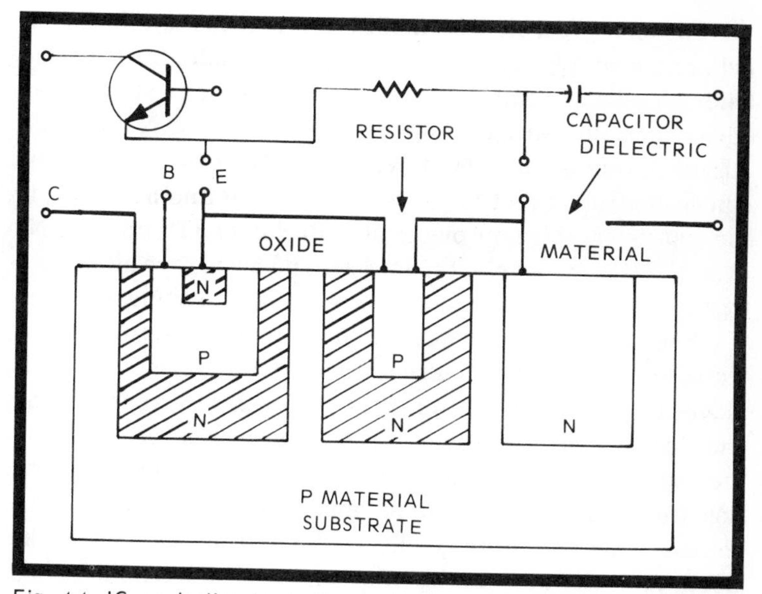

Fig. 4-4. IC production is similar to the technique used for transistors. Resistors and capacitors can also be formed. The schematic (top) is the equivalent circuit.

In the field, the troubleshooter who encounters all types of electronic gear cannot set up to test an individual chip. Besides that, ordinary schematic diagrams do not give the "closeup" schematic, but instead a triangle, with a number like CA3020, and a dozen leads. How can such a chip be tested?

The best way is to have a spare chip, like a replacement tube, and install it. The direct substitution gives the very best test. However, this is not always possible. Therefore, the only way left is an input-output test. For instance, suppose a chip is used as the audio circuit. A test tone in the audio range like 1-kHz is taken from a signal generator. It is applied to the speaker. If a signal is heard, the speaker is good. The next test point is the center tap of the volume control. The 1-kHz signal is injected there. If no audio is heard, the circuit between the speaker and the control is defective, in this case it is the chip.

Similar tests can be made in other circuits with the usual troubleshooting techniques. The circuits are identical in a chip, they are just microscopic.

SILICON CONTROLLED RECTIFIERS

The term, rectifier, in the SCR name tends to mislead. It indicates a simple type device, which is actually far from the truth. A silicon controlled rectifier is the next step up as devices are built with the building blocks of P and N material. Just as the transistor is built by adding a block to the diode, the silicon controlled rectifier is built by adding another block to the transistor. It is four pieces of material, two Ps and two Ns in an NPNP arrangement (Fig. 4-5). The pieces are all bonded together, forming three junctions where the four pieces meet.

You can consider an SCR as three diodes in series or as a transistor and diode in series. The forward bias, as usual, places a positive charge (or a deficiency of electrons) on the end P piece and a negative charge (or excess of electrons) on the end N piece. When the device is forward biased, electrons flow from the negative terminal and through the bottom three pieces, which are exactly like an NPN transistor. Then the electrons easily pass through the last piece of material just as they would pass through any forward biased diode. Therefore, in forward bias, electrons pass through the device easily.

When the device is reverse biased, electrons try to flow the other way. However, the electrons cannot get past the piece of P material in the top diode. Therefore, in reverse bias the device acts just like an ordinary diode rectifier.

The piece of P material is conventionally called the anode. The bottom piece of N material is also conventionally named the cathode. However, a third lead is connected into the bottom piece of P material (Fig. 4-6). It is called the gate, like the name given to FETs middle electrode.

With forward bias applied, the three junctions are charged. The bottom most junction has an excess of electrons coming in through the cathode. The electrons leap the junction and the junction becomes forward biased (Fig. 4-7). As the electrons reach the second junction they encounter a piece of N material on the other side which has a repelling effect. That makes the middle junction reverse biased.

Meanwhile, the deficiency of electrons on the anode draws electrons out of the top piece of P material. This places a deficiency of electrons on the P side of the top junction.

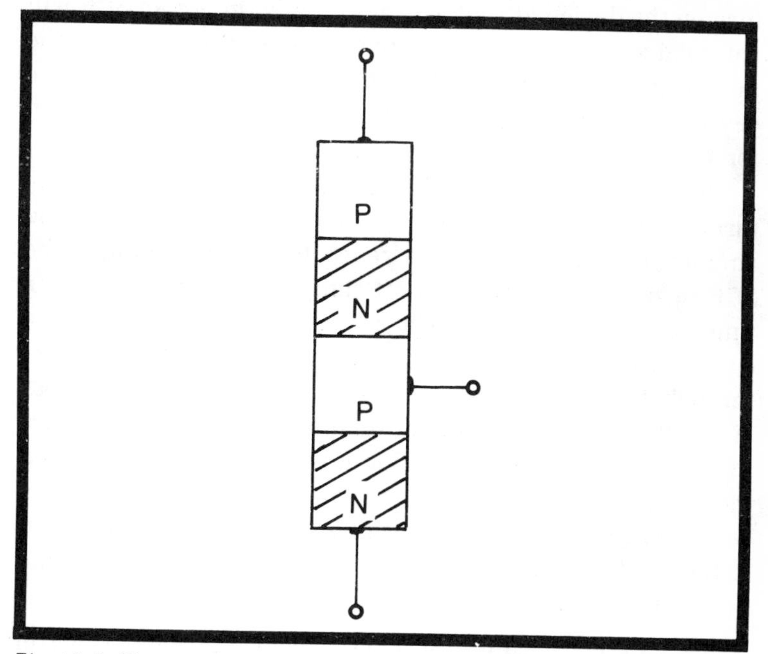

Fig. 4-5. A silicon controlled rectifier is formed by adding another piece of N or P material to a transistor.

Electrons then flow from across the junction piece of N material. The top junction is thus forward biased.

Therefore, electrons pass easily from the cathode, across the forward biased bottom junction, across the reverse biased middle junction (by transistor action), then across the forward biased top junction on to the anode. However, the gate is attached to the bottom piece of P material and has a large influence on the naturally reverse biased middle junction. If a negative voltage is applied to the gate, it increases the number of electrons that get to the P side of the middle junction. This widens the junction or increases the amount of reverse bias. If the negative voltage is high enough, the current from cathode to anode is cut off.

On the other hand, if a positive voltage is applied to the gate, it takes off some of the excess electrons on the P side of the middle junction, making the junction narrower and thus forward biased. The electron flow through the device passes the narrower middle junction easier and the device turns on.

A gate pulse can be applied to a negative biased gate and turn on the device at each pulse. Actually, in use an SCR will not conduct. It needs the positive gate bias to conduct. The middle PN junction with its natural reverse bias keeps the device off. A **forward breakover** voltage on the gate is needed for the turn on.

The gate current controls the amount of voltage needed to turn on the SCR. The more gate current the smaller the amount of voltage needed to reach breakover voltage. The SCR turns on fully and shuts off completely. It is valuable in time delay circuits and power control. It can turn on a large current with a low power loss. Very little energy is dissipated through it. For instance, 500 watts can be controlled with a loss of two watts. Consider this in relationship to potentiometers that control circuits and dissipate hundreds of watts during the control. SCR devices are available in various voltage and wattage ratings. The reverse voltage and wattage is identical to the top PN junction rating. If it breaks through there, the device is ruined.

SCRs are also called thyristors because they have control characteristics like a thyratron tube. The thyristor name is losing, though, to the term, SCR. SCRs are finding their way into many circuits and will replace tubes and transistors in applications like low-frequency oscillator circuits. The completeness of the off and on modes approaches a square wave. This enables subsequent circuits to shape the all purpose square wave into the exact wave desired.

The schematic symbol representing an SCR is a diode in a circle with the anode and cathode in conventional positions. Attached to the cathode is a straight line to one side. It represents the gate (Fig. 4-6). The diagram simplifies the complex activity that is taking place inside the SCR.

Testing the SCR

There is no easy SCR test due to the inaccessibility of the top junction. Quick checks with an ohmmeter provide only one test. It shows up a shorted junction. If the top junction shorts through, an ohmmeter reading between the anode and gate will resemble that of an ordinary diode; that is, a low resistance will show on the ohmmeter when you attach the probes so that the middle junction is forward biased. When you reverse the probes the ohmmeter then reads a high resistance.

For an ordinary diode, those results would mean the diode is good. However, since there is an extra junction, (the top one) between the gate and the anode, a good SCR will read a high resistance in both directions because the extra junction is opposite to the middle junction. A quick check between the cathode and gate also reads like an ordinary diode, since there is only one junction—the bottom one between the two elements.

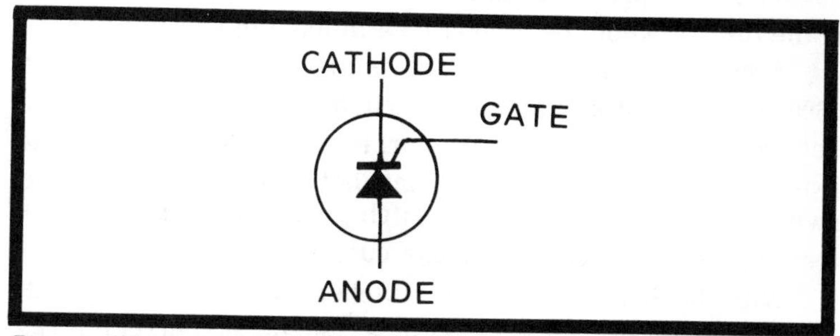

Fig. 4-6. Schematic symbol of an SCR. Terminals are the cathode, gate and anode.

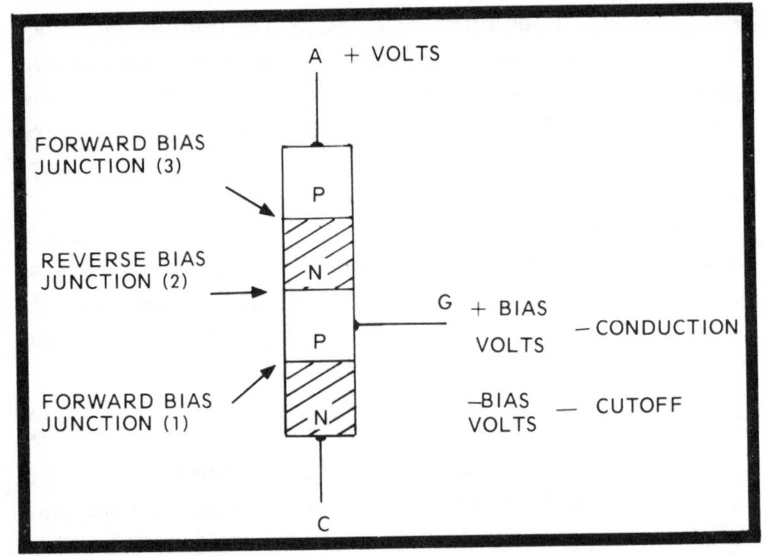

Fig. 4-7. The electrons from an SCR cathode pass through junction 2 by transistor action during forward gate bias. During reverse gate bias junction 2 cuts off the flow.

To sum up, if the diode resistance test between anode and gate reads like a good diode, then the tested section has a shorted junction. If you try the diode resistance test between cathode and gate, then the diode resistance test is valid.

Unfortunately, the diode resistance test does not reveal when the top or middle junction is open. This almost makes the preceding resistance test a waste of time, for it only tests the top and middle junctions for shorts. Also, most transistor testers do not provide an SCR test. So, what can be done? RCA recommends a small circuit (Fig. 4-8) that gives a good go no-go test for opens and shorts. Since SCRs fail that way most of the time, it would be a good idea to have the circuit handy.

A 25-watt ordinary light bulb is attached in series with a 5600-ohm resistor and an on-off switch. A test lead, connected between the light bulb and resistor, is the anode lead. The gate lead is attached to the center tap between the resistor and switch. The other end of the switch goes to the cathode. The test source is the 117-volt house current.

With the SCR to be tested attached, the switch is closed. Due to the total circuit resistance, 20 ma of current flows through the light bulb, resistor and switch, from the source and back to the source. This is not enough current to heat up the filament in the light bulb and cause light. Therefore, the light is out due to the switch short across the gate-cathode part of the SCR. Now, the actual test is made by opening the switch.

One of three things will happen. The lamp will light dimly, brightly or not light at all. If it lights dimly, the SCR is good. The SCR gate gets a 30 ma positive pulse on each positive cycle of the line current. With the switch open, the line current is passing into the gate-cathode area of the SCR. On the positive cycle the gate-anode area of the SCR becomes a low resistance, shunting the 5600-ohm resistor and allowing the bulb to draw more current from the line, since it is the only appreciable amount of resistance left in the series circuit.

Should the lamp light brightly, there is no resistance left in the SCR and the lamp draws full current from the line. The SCR is shorted through. According to the nature of the short, the lamp could light brightly whether the switch is off or on. If the lamp will not light when the switch is either open or closed, the SCR is open.

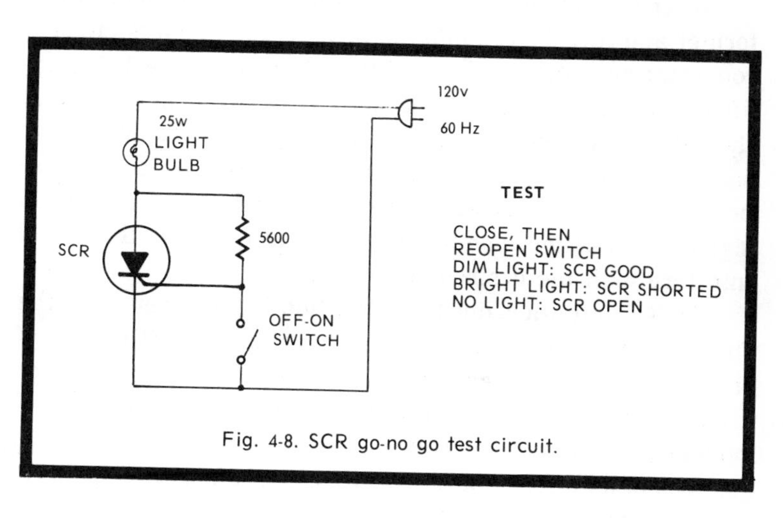

Fig. 4-8. SCR go-no go test circuit.

CHAPTER 5

RF Amplifier Circuits

The radio frequency or RF stage in a receiver is not necessarily an amplifier. It can simply be a tuned stage with no amplifying device. Without that device the stage is identical in either tube or solid-state instruments. When the RF stage amplifies, though, it has either a tube or transistor. Both circuits are quite similar except for the active device.

A typical RF stage has two tuned parts—input and output. The input is the antenna and is tuned by the physical dimensions of the antenna and an adjustable coil or capacitor that transfers the incoming signal from the antenna to the amplifying device (Fig. 5-1).

After the signal is amplified, it is fed into a tuned transformer and transferred into the next stage, which is either a converter or another RF amplifier (Fig. 5-2). The RF amplifier provides excellent isolation between the antenna and the rest of the receiver circuits. It keeps a lot of interfering signals from getting into the receiver. Signals that cause problems are electrical interference and other generated signals such as those on image and IF frequencies that could interfere with the desired reception.

An RF amplifier, due to its location in the receiver as the first amplifier, must have certain characteristics (Fig. 5-3). If it loses these characteristics, it will amplify noise and other troublesome signals. Whatever the RF stage amplifies will appear greatly magnified in the output.

The number one characteristic is sufficient stage gain. It has to be around 100. If it's considerably less than 100, it does not provide enough "punch" to the incoming signal. If it's a lot more than 100 it's liable to amplify static and internal amplifier noise which will also be displayed in the output at an undesirable level. About 100 gives a middle ground amount of amplification, resulting in a good signal-to-noise ratio.

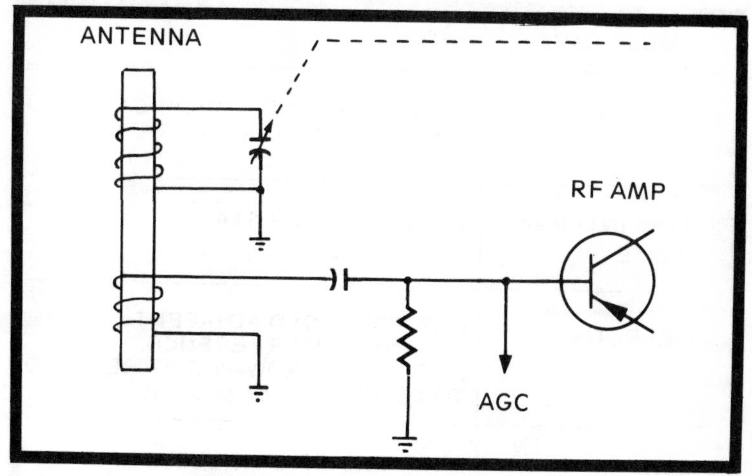

Fig. 5-1. A typical RF amplifier has a tuned input ganged with the output. The antenna is part of the resonant circuit.

The second characteristic is a low level of internal noise in the amplifying device. Since the RF amplifier is the first in the receiver, a lot of internal noise in the stage output will be amplified along with the incoming signal throughout the receiver. Such noise appears as snow in a TV picture or static from a speaker.

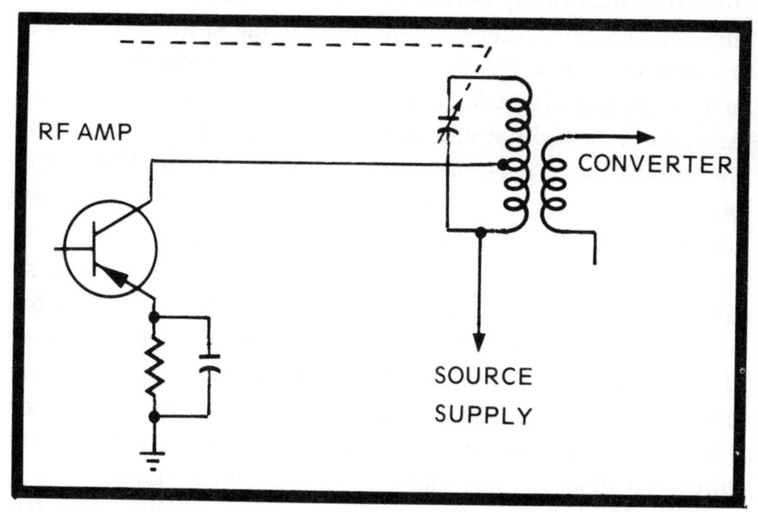

Fig. 5-2. The RF output tuning is ganged to the input tuning and it couples the signal to a converter stage. Notice the stepdown transformer.

MUST HAVE	WHY?
1—GAIN, ABOUT 100	MIDDLE GROUND AMPLIFICATION; GOOD SIGNAL TO NOISE RATIO
2—LOW INTERNAL NOISE	CAUSE SNOW OR STATIC
3—SELECTIVITY	NOT TOO BROAD-ADJACENT CHANNEL INTERFERENCE. NOT TOO NARROW—WON'T RESPOND TO COLOR TV OR FM STEREO.
4—LINEAR	HIGH END OF DIAL JUST AS GOOD AS LOW END
5—ISOLATION	KEEP LOCAL OSCILLATOR FROM RADIATING

Fig. 5-3. A good RF amplifier must have these five characteristics in order to perform properly.

The third characteristic is selectivity. The RF amplifier must tune accurately across the range of the receiver. It can't tune too broadly; otherwise, more than one station signal will be received. In a TV this could cause adjacent-channel interference and in a radio, a number of stations could be heard. Also, the RF amplifier can't be too selective or it could narrow the pickup of a particular transmission. In an FM stereo broadcast, the stereo effect would be missing if the stage were too selective. In a color TV, the color could be missing.

The fourth characteristic is linear amplification. Since an RF amplifier tunes across a broad range (and it is impossible to tune exactly the same throughout), it must have linear characteristics, so that even though it can't be perfectly linear, it is linear enough for satisfactory output. In other words, stations received at the high end of the dial should be amplified about the same amount as those on the low end.

The fifth characteristic is its isolation ability. The RF amplifier must prevent radiation generated by the oscillator in the converter stage from getting back into the antenna. This type of radiation could play havoc with nearby receivers. A good RF stage successfully isolates the antenna from the local oscillator.

BIPOLAR RF AMPLIFIERS

Bipolar transistors (the ordinary transistors) are commonly used in all types of receivers. A typical RF stage in a TV or FM receiver employs a PNP transistor in what is known as a common-emitter configuration (Fig. 5-4). By common emitter, it is meant that the emitter is near ground almost identical to the way a cathode in a tube circuit is tied near ground. Actually, the emitter can be soldered directly to ground or there can be a bias resistor between the emitter and ground. When the bias resistor is used, it is normally bypassed

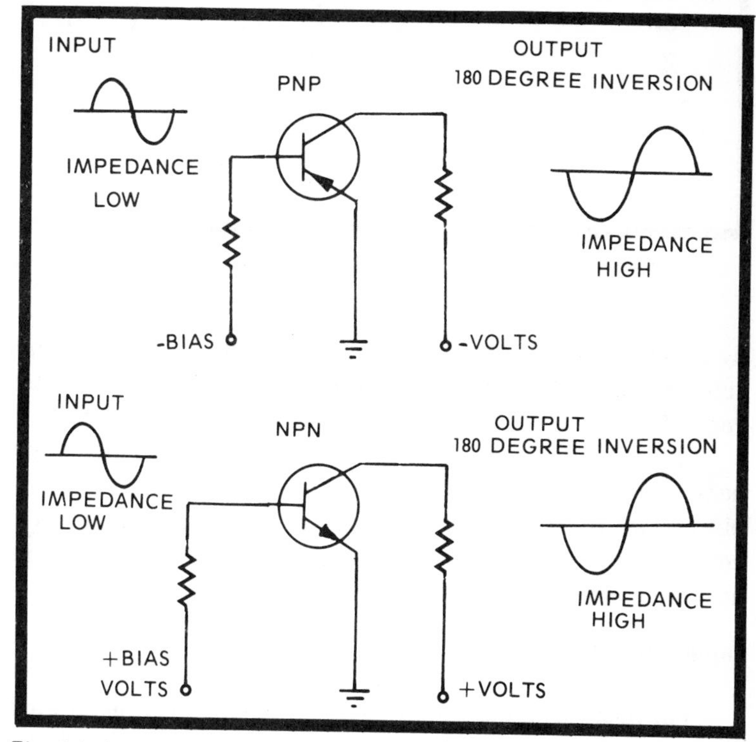

Fig. 5-4. A common-emitter amplifier produces high current, voltage and power gains.

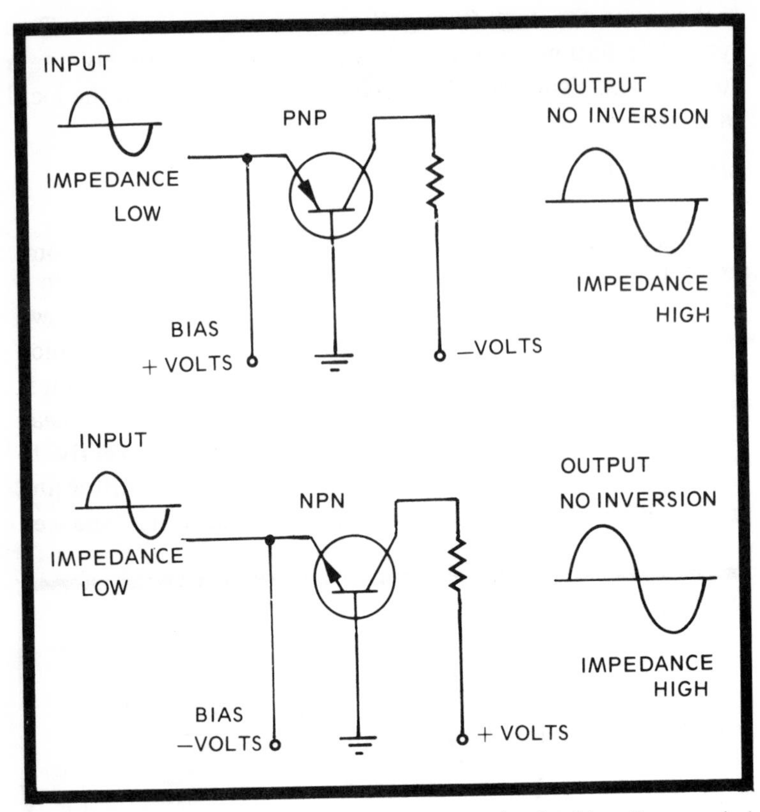

Fig. 5-5. A common-base circuit produces a current gain of less than one, but high voltage and medium power gains.

by a small capacitor connected across the resistor to stop any signal from being developed across the resistor. With the capacitor, only a constant DC bias is placed on the emitter.

The common-emitter circuit, just like the grounded (or nearly so) cathode, is popular because maximum amplification is obtained. Other transistor configurations—common-base and common-collector— are not as efficient in amplifier applications (Figs. 5-5 and 5-6).

Input Circuit

The base circuit is usually quite extensive. It must match the antenna to the transistor, trap out image and IF frequencies and also tune in the desired signal. In TV and FM

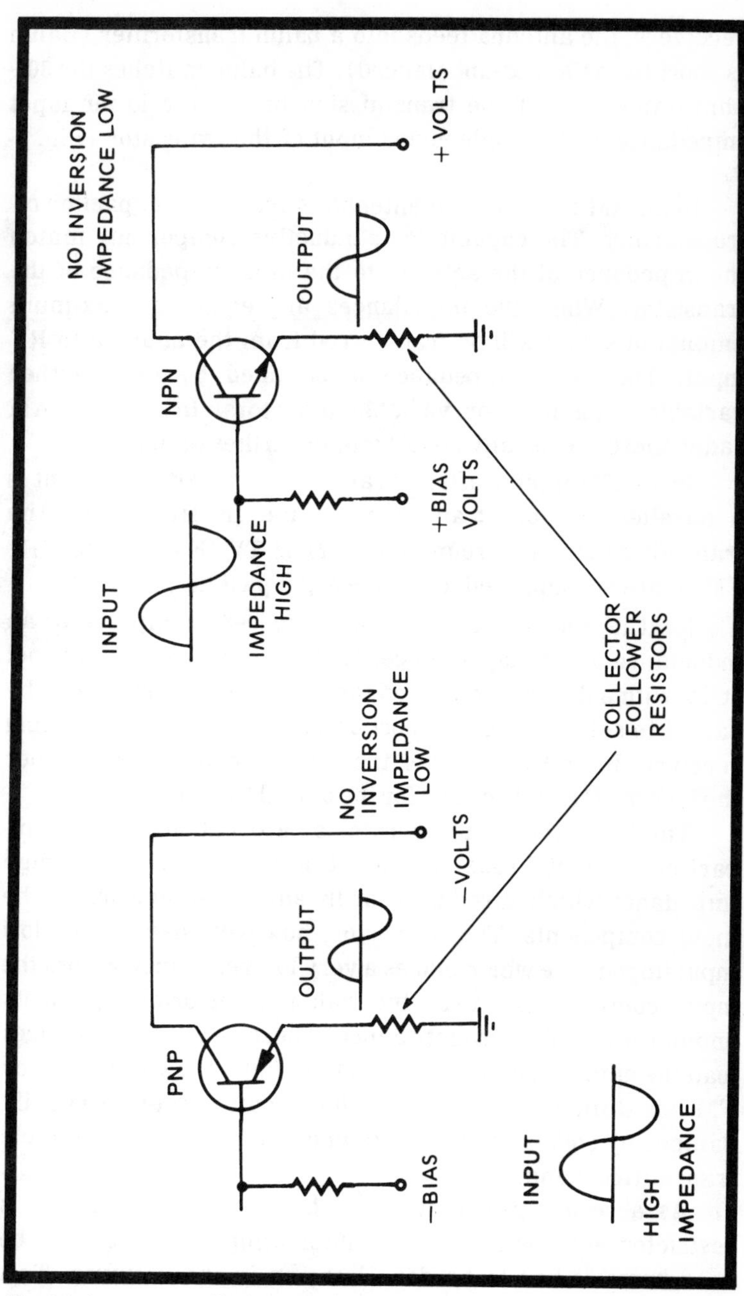

Fig. 5-6. A common-collector circuit produces a voltage gain less than one, but a high current gain and medium power gain. It operates like a cathode follower tube circuit.

receivers, the antenna feeds into a balun transformer (Balun is short for balanced-unbalanced). The balun matches the 300-ohm impedance of the transmission line to the lower input impedance of the single-ended input of the transistor (Fig. 5-7).

In an AM receiver the antenna is fed into a capacitor or transformer. The capacitive or inductive components match the impedance of the antenna to the input impedance of the transistor. When the impedances are equal, a maximum amount of signal will be transferred from the antenna to RF input. The exact impedance is obtained by using either variable capacitors or variable inductors. In a small AM radio, there are usually no IF traps or further tuning.

In TV-FM inputs, after the antenna transfer component is a parallel resonant trap tuned to the IF frequency. Any transmitted signals around the 44-MHz TV IF or the 10-MHz FM IF are not supposed to be able to get past that trap.

Next in line is a tuned circuit (Fig. 5-8) composed of an inductance and a capacitance. In a TV the L is the channel strip, with a different one used for each channel. The C can be varied to select the exact portion of the signal that is being received. In an FM receiver the L is not changed, but either the L or the C could be used for the actual tuning.

The transistor input operates at a disadvantage in comparison to an RF tube. The input of a tube has a very high impedance which introduces hardly any resistance across the input components. The transistor, however, has a very low input impedance which places a very low resistance across the input components. Also, the balun transformer presents another lowered input impedance. These two low resistances load the entire circuit heavily and the full bandwidth in TV or FM has difficulty in passing. The circuit has to be heavily damped by coils and capacitors in order to pass the complete transmitted band.

As a result, selectivity in a transistor input circuit is restricted and the rejection of interfering signals cannot be used here; it has to be done later on in the receiver. This resulting wide-band reception lets a lot of different signals into the transistor RF and they are all amplified. However,

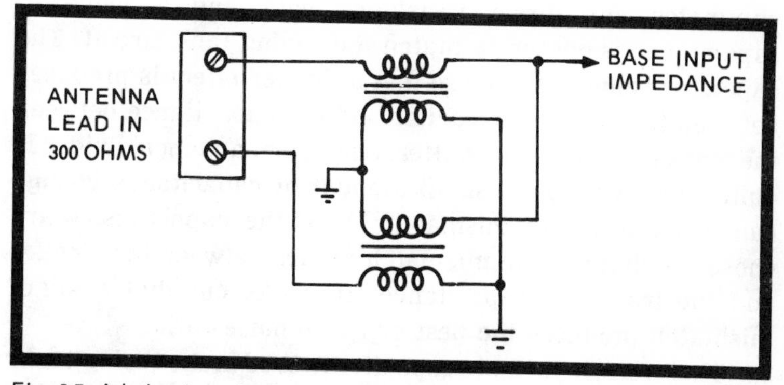

Fig. 5-7. A balun transformer matches the 300-ohm transmission line to the base of the RF amplifier in a TV set or FM radio.

transistor circuits do achieve a high signal-to-noise ratio and the rest of the selectivity is postponed.

It might be argued that a tube RF amplifier should be used if the bipolar transistor is not as good. Possibly; however, a tube socket is needed, heater voltages are needed and the portability and miniaturization possible with transistors would be lost. It works out on balance that even with the hardship of excessive loading, the transistor still has more advantages. In the next section on FET RF amplifiers, it is shown that this difficulty is overcome, since the FET has the same type of high impedance input as the tube, yet still retains all the advantages of the transistor.

In the tuning part of the bipolar transistor RF input, the L is tapped and thus matches the low input impedance of the antenna. Then, a blocking capacitor is placed between the tuning section and the base of the transistor. The capacitor

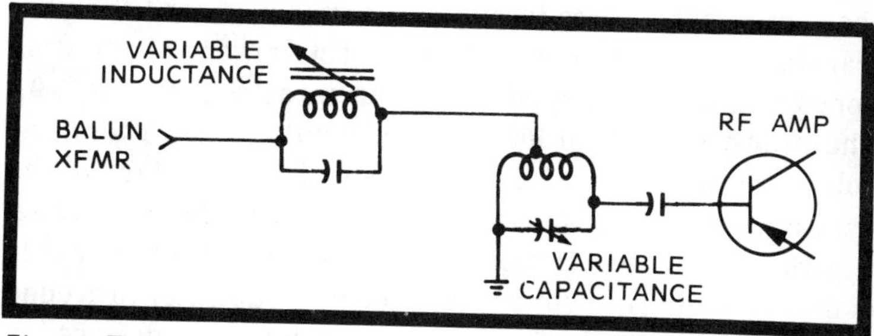

Fig. 5-8. The input circuit is tuned by a variable inductance and a variable capacitance.

eliminates any direct resistance path and a value of capacitance is chosen to match the tuning tank circuit. The capacitor value is chosen so that a divider effect is produced between the blocking capacitor and the capacitance between the transistor base and emitter. Such a capacitance divider is quite sensitive; just a small amount of capacitance change can cause a severe mismatch. Also, the capacitances are chosen so that a slight mismatch occurs between the blocker and the transistor capacitance. It works out that a slight mismatch produces the best signal-to-noise ratio.

Therefore, a variable capacitor is used in the tuning tank. This capacitor is part of the base circuit. It is a large value capacitor so that in ratio it makes the transistor capacitance almost insignificant. The entire base capacitance circuit can, therefore, be tuned exactly with the tank capacitor for best signal-to-noise ratio. If the RF transistor or any of its input components are changed, the tank capacitor has to be touched up to compensate for changes in transistor input and output capacitance.

Output Circuit

The signal enters the transistor base. Electrons leave the base and go to the emitter of the PNP transistor. The electron flow is varied by the signal. Electrons are also leaving the collector and arriving at the emitter. The B to E flow causes the C to E flow to vary in exactly the same manner. Since the resistance between C and E is many, many times larger than the resistance between B and E, a larger voltage drop takes place in the C to E flow.

The electrons, originating in the minus voltage section of the power supply, pass through the primary of the RF output transformer and then from C to E. The larger voltage drop appears in the primary of the output transformer (Fig. 5-9). The large drop is about 100 times the input voltage. Therefore, voltage amplification has taken place. The secondary of the transformer, of course, couples the signal to the next stage.

The transformer is designed to tune throughout the range of the receiver. The primary is tuned by a capacitor to ground in the collector circuit. The Q of the transformer is designed to pass the required band but reject the immediate adjacent signals. In a TV, that would mean the transformer would pass the full 6-MHz passband and in an FM receiver the full 150 kHz. There are, of course, some TVs and FM radios that are intended to pass less signal for economy as well as other reasons.

Neutralization

A major problem in an RF amplifier is its ability to start oscillating on its own. Due to the frequencies that must be processed and the steps taken to attain large amounts of gain, this type of parasitic oscillation can break out. In all active components, such as transistors, there are small amounts of capacitance between elements. There is capacitance between E and B, between B and C and between E and C. There is also a small amount of inductance between the same junctions. Admittedly, the amounts are small but large enough to produce impedances that will "act up" under conditions of high frequency and large gain.

Such impedances can act as tiny tank circuits inside the transistor, and while the transistor is amplifying the signal, the unwanted distributed capacitance can feed a bit of the signal back from the collector to the base. This causes the RF stage to start oscillating.

It becomes obvious that if we reduce either the frequency or the gain of the stage, the oscillation tendency will drop. Since the frequency can't be lowered, that leaves the gain. Reducing the gain is one method of getting rid of this an-

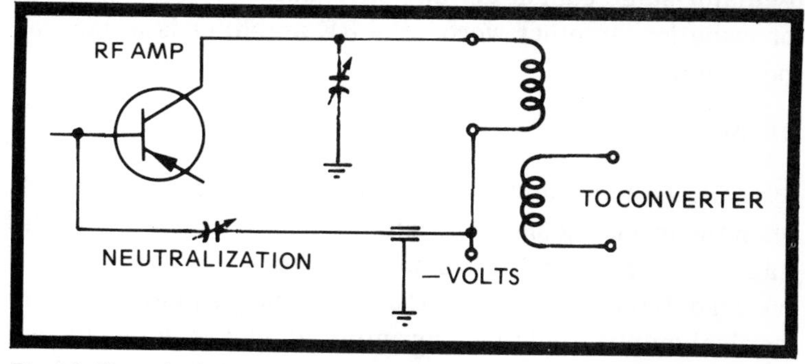

Fig. 5-9. The output circuit adds to the selectivity and couples the signal to the following stage. Neutralization is accomplished with a variable feedback capacitor from the output to the base.

noyance. When the gain is lowered, the RF amplifier is not quite as effective but at least it doesn't oscillate. In strong signal areas or if two stages of RF are used, the gain is not missed. However, in weaker signal areas, it would be advantageous to step up the gain.

There's another method of damping the oscillation—it's called neutralization. Since the collector is being tapped by an invisible capacitor which is feeding signal back to the base, 180 degrees out of phase, all we have to do is take some other signal and feed it in phase to the base. That way the in-phase and out-of-phase signals cancel or neutralize each other.

On the other side of the output primary is a signal that is completely out of phase with the feedback. A tap is made there and a variable capacitor installed. The output of the capacitor is fed directly to the base. This neutralizing signal is also the same phase as the incoming RF. The incoming RF is changed 180 degrees as it passes from base to collector. It is then inverted again as it passes from the top of the primary to the bottom. This signal then reinforces the weakening RF as well as cancels the degenerative feedback. The reinforcement helps a tiny bit in improving the signal-to-noise ratio also.

The capacitor, called the neutralizing capacitor, is tuned until it produces the best output. At that point the neutralizing voltage exactly equals the feedback voltage. The collector capacitance is thus eliminated. The input and output circuits become completely isolated from one another. The RF signal is amplified smoothly with less amounts of noise and the oscillator stage cannot get any of its output into the antenna for radiation. In other words, the RF amplifier is a one-way thoroughfare.

RF AGC

AGC—automatic gain control—is needed in an RF stage when the incoming signal becomes too large. The gain of the stage must be "braked" somewhat so the stage will not overload. If there were no AGC during strong signal reception, the final output would be overamplified and sound distorted. During a weak signal input the AGC is hardly used. If all signals were weak, the RF AGC could be dispensed with.

The AGC is used, though, even with weak signal reception, but primarily for correctly setting the DC bias of the base. When a strong signal comes in, then it, in addition to setting bias, also reduces the gain of the stage.

In tube receivers the RF stage normally receives a negative bias. That is the only type used in such circuits. In a transistor RF stage, two different types of AGC could possibly be used. One is reverse bias and the other is forward bias.

Reverse Bias

Reverse bias in an RF stage is similar to tube-type bias. In a PNP RF transistor, as a signal gets stronger a positive-going voltage is applied to the base. Since a transistor is forward biased when the emitter is more positive than the base, the addition of a positve voltage on the base tends to make the base more positive. This reduces the amount of bias on the EB junction. Reduction of the bias there reduces the electron flow from base to emitter. If the bias is made positive enough it will make the base voltage as positive as the emitter voltage and the electron flow will then halt.

The reduction in the base-to-emitter electron flow causes the collector-to-emitter electron flow to act in the same way. Therefore, the reverse bias can reduce the gain of the amplifier. However, as the DC operating point lowers, the input and output impedances of the transistor change. This mismatches the RF stage and the gain is reduced even more.

As the signal gets weaker, though, the reverse bias is not as strong and the transistor base approaches normal DC operation. This enables the collector current to increase, which makes the beta of the device increase, thus compensating for the weakness of the signal.

Forward Bias

While reverse bias keeps the RF amplifier from overloading by changing the E to C electron flow, or to put it another way controls collector current, forward bias controls the collector voltage (Fig. 5-10). The AGC still is applied to the base, but instead of applying a voltage that tends to reduce the E to B bias as before, the applied voltage tends to increase the forward bias.

In a PNP transistor the E to B bias holds the emitter more positive than the base. This causes current flow from the base to the emitter. If a negative-going voltage is applied to the base, the emitter becomes even more positive and more electrons will flow to it. This also causes the collector to send more electrons to the emitter. Therefore, a forward bias will make the RF amplifier conduct in a heavier manner. How can this cause a collector voltage drop?

In forward bias an extra resistor is inserted in series with the collector. The resistor is bypassed with a capacitor to ground so no signal can develop across it. As the collector current increases due to the forward bias, more electrons are dragged through the resistor and the increase in current causes a larger voltage drop across the resistor. Therefore, the collector voltage decreases. The voltage drop will vary in accordance to the incoming base AGC.

Since amplification is due to the difference in resistance between the EB and EC junctions, the lowered collector voltage causes less signal to develop across the EC junction. To sum up, the larger the current flow from E to B, the lower the signal voltage developed across the E to C junction.

The foregoing RF amplifier, using a bipolar transistor, is commonly used. But there are simpler ways of getting the function performed. The simplicity is achieved in the coupling from the antenna to the base of the transistor. The rest of the circuit—output, neutralization and AGC—remain much the same. The simpler circuit is used mostly in AM receivers. Instead of the previously described coupling, a simple capacitive or inductive coupling can be used.

The main duty of the coupling circuit is to match the antenna output to the transistor input, at least close enough for all the stations to be tuned. The impedance of the antenna, when matched to the low input impedance of the base circuit, transfers the maximum amount of signal.

Capacitive coupling is accomplished with a simple circuit. A network consisting of two capacitors in parallel with a small inductance between them does the trick neatly. With the correct values of capacitance, the match is easily made.

Inductive coupling is accomplished also with a small circuit. An antenna transformer with a primary and a secondary does the impedance matching. The primary has enough turns to exactly match the antenna impedance. The secondary has the number of turns required to exactly match the base impedance.

FIELD-EFFECT TRANSISTOR RF AMPLIFIERS

A field-effect transistor offers many advantages as an RF amplifier over the bipolar transistor. The FET has practically no loading effect on the input, since its shunt resistance is very high. In contrast, the base input of a bipolar transistor is lower in resistance than the antenna.

Since the FET input is so high, the FET can be connected across the entire antenna coil with little or no thought to the impedance match. Then, the antenna can be attached to a tap on the coil. The antenna coil becomes a stepup autotransformer with a subsequent increase in gain. Old reliable tube-style circuits can be used with an FET. The FET has characteristics very similar to a tube without the need for a heater supply or the associated tube components.

An example of an old tube-type RF amplifier is the cascade circuit, using two triodes, one on top of another

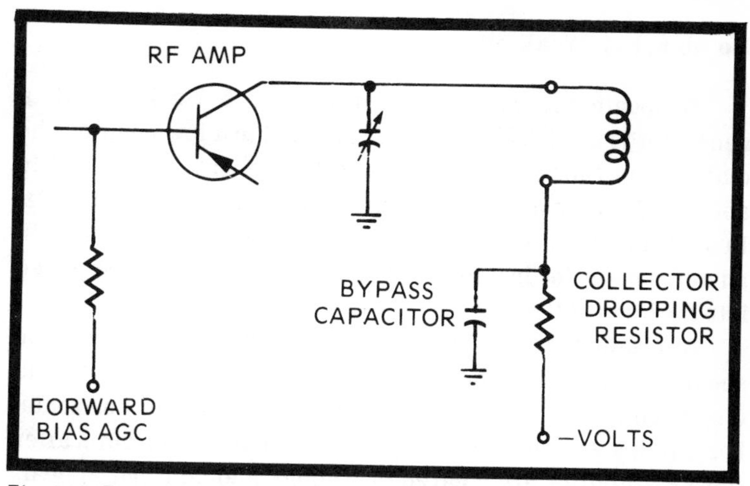

Fig. 5-10. Forward bias AGC reduces the collector voltage by dropping additional voltage across a special, bypassed load resistor.

(schematically speaking). The cathode of the top one is attached to the plate of the bottom one. This gives the circuit the gain of a pentode and the noise free characteristics of a triode. The dual-triode is contained in one glass envelope. A typical version popular several years ago was the 6BQ7.

Now the same circuit can be used with a dual-gate MOSFET. The electrons flow through the channel from source to drain, or drain to source, according to the polarity of the channel. The RF signal enters through gate 1, just like the RF signal enters the cascade bottom control grid. The second gate is biased to pass the signal. The amplified, relatively noise-free signal appears at the drain and is coupled to the next stage in the tuner. The AGC voltage can be applied at gate 1. RF signal matching is simple.

TESTING RF CIRCUITS

An RF amplifier is a fairly straightforward amplifier. The inputs are the antenna feed and AGC. The output is the amplified version of the signal input. The RF amplifier can decrease in gain or lose the signal altogether, or it can cause signal overloading in following stages. The first step in testing any circuit is the localization of that circuit as the troublemaker.

No Signal or Weak Signal

To locate the cause of no signal, or a weak signal, a generator is used. The generator is tuned to a low point on the dial. For instance, in an FM radio the generator is tuned to about 88 MHz. In a TV the low point is Channel 2 or 55.25 MHz. In an AM radio it is 560 kHz. The test signal is modulated with an audio note; for instance, 400 Hz. Then the signal is injected into the input of the stage after the RF amplifier.

In a radio the modulation will be detected and heard in the speaker. In a TV the modulation will be detected and displayed in the TV picture as a number of black horizontal bars. If this detected signal appears, then all the circuits after the RF amplifier are cleared of trouble. Should it not appear, the RF amplifier is probably good and the seat of the trouble is further on. When it does appear, however, the injection point becomes the input of the RF amplifier. If there is no signal now, the RF amplifier is indicated as the circuit in trouble.

Signal Overload

When the symptom in the receiver is signal overload, it could mean that the AGC is not getting to the RF amplifier. If so, the test to prove the point one way or the other is easy. Take a bias box and attach it to the AGC input of the RF amplifier. This means on the AGC side of the isolating resistor in the AGC line. Adjust the bias box for various DC bias voltages.

If the overloading clears due to the external bias, the AGC is not arriving at the RF amplifier and there is trouble in the AGC line. When the AGC substitution does not help the overloading, the AGC is probably good and the condition is being caused in the RF circuit itself. (This test is performed after the IF section is exonerated as discussed at the end of Chapter 6.) Further checks in the RF section must be conducted.

Once the RF amplifier is identified as the circuit in trouble, routine voltage and resistance readings are made. The actual component is pinpointed and replaced.

TV RF ALIGNMENT

TV RF alignment procedures vary so much from TV to TV that a set of universal instructions cannot be laid out. You must have the manufacturer's alignment notes and follow them (Fig. 5-11). The only general rules are that the RF tuner response should be 6MHz with a peak of gain at the middle of the channel. Both the sound and picture carriers should be centered halfway up on the two slopes.

Such a curve is for the tuner response only. Any overall RF-IF response check must be made after the IF has been aligned. The RF-IF overall response curve is almost identical to the IF overall response curve. See the next chapter on IF alignment.

PRELIMINARY INSTRUCTIONS (READ CAREFULLY)

1. Keep all exposed "hot" leads as short as possible, minimum length precautions also apply to all ground leads including braid and shielding.
2. Use shielded leads between all test equipment and chassis.
3. Terminate all ground leads and lead shields as close as possible to their respective "hot" leads.
4. Use high scope gain and keep sweep generator output at lowest usable value; check at intervals for possible sweep generator over-loading by temporarily varying signal input level and noting any change (excluding amplitude) in response curve shape.
5. Keep marker generator coupling to a minimum to avoid distortion of response curve.
6. For optimum receiver alignment, power line voltage should be maintained at 120 volts.
7. Use non-metallic tools for all alignment adjustments.
8. Connect +3V to AGC terminal.
9. Connect +5.6V DC to AFC terminal.

VHF OSCILLATOR ALIGNMENT

This tuner employs a pre-set oscillator adjustment screw for each channel. No other adjustment is necessary other than the normal fine tuning adjustment. If all available stations cannot be received and fine tuning range is not adequate, adjust aluminum core of L34 as per steps below.

1. Set fine tuning control fully clockwise.
2. Set IF marker generator to 41.25MHz.
3. Sweep generator sweeping 10MHz minimum.
4. Terminate IF output with 75 OHMS.
5. Connect tuner as shown in Fig. 1.

TUNER SETTING	SWEEP GENERATOR	RF MARKER GENERATOR	ADJUST
Ch. 13	213MHz	211.25 MHz	L34 until 41.25MHz marker coincides with 211.25MHz marker.

6. Set Fine Tuning to mid-range.

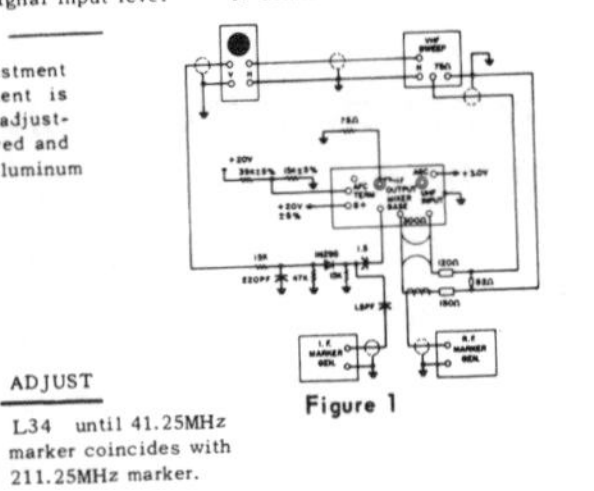

Figure 1

45.75MHz

NOTE: Neglect response shape if due to the loading of IF marker generator.

RF ALIGNMENT

Detune mixer collector coil and rotate tuner, sweep generator and RF marker generator through all channels, refer to "Alignment Frequency Chart" for frequencies of specific channels, Should picture or sound marker fall below 50% or any channel, the tuner must be realigned as follows:

If proper response curve cannot be obtained, adjust C12 with tuner set to channel 7 as follows:

Set up tuner as described in Fig. 2 using detector "B" at IF output. Feed a strong channel 7 sweep signal into the antenna terminals and adjust C12 for a null in the response.

Next, connect detector "A" to mixer base as illustrated in Fig. 2 and return AGC to +3V.

Set fine tuning to midrange and perform the following:

S P

20 MV

Figure 2

TUNER SETTING	SWEEP GENERATOR	RF MARKER GENERATOR	ADJUST
Ch. 13	213MHz	215.75MHz 211.25MHz	L17, L28, LA, LB for critically coupled response curve. L7 for symmetrical response curve.
Ch. 6	85MHz	83.25MHz 87.75MHz	Spacing between and the inductance of L310 and L210 for critically coupled response curve. L510 for symmetrical response curve.
Ch. 5	79MHz	81.75MHz 77.25MHz	Same as Ch. 6 using coils L308, L208. L508 for symmetrical response curve.
Ch. 4	69MHz	71.75MHz 67.25MHz	Same as Ch. 6 using coils L306, L206. L506 for symmetrical response curve.
Ch. 3	63MHz	65.75MHz 61.25MHz	Same as Ch. 6 using coils L304, L204. L504 for symmetrical response curve.

VHF TUNER ALIGNMENT PROCEDURE (CONTINUED)

RF ALIGNMENT

TUNER SETTING	SWEEP GENERATOR	RF MARKER GENERATOR	ADJUST
Ch. 2	57MHz	59.75MHz 55.25MHz	Same as Ch. 6 using coils L302, L202. L502 for symmetrical response curve.
NOTE: Readjust tuner IF output coil L22 for proper response curve as shown on pg. 14, Figure 3 before operating tuner with main chassis.			
Ch. 1	43MHz	45.75MHz 42.25MHz	Using Channel 1 simulator shown in Figure 4 and detector "A" connected as shown in Figure 2, increase AGC voltage to tuner to provide 10DB less gain. Then adjust L300 and L200 for critically coupled response curve. Adjust for a symmetrical curve using coil L13 .

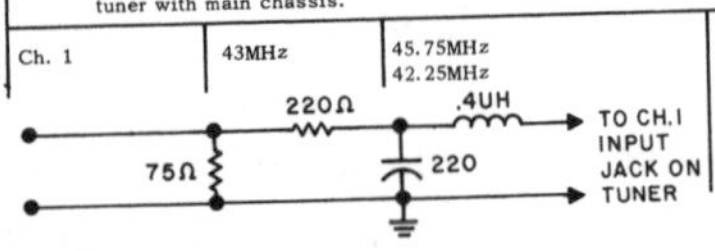

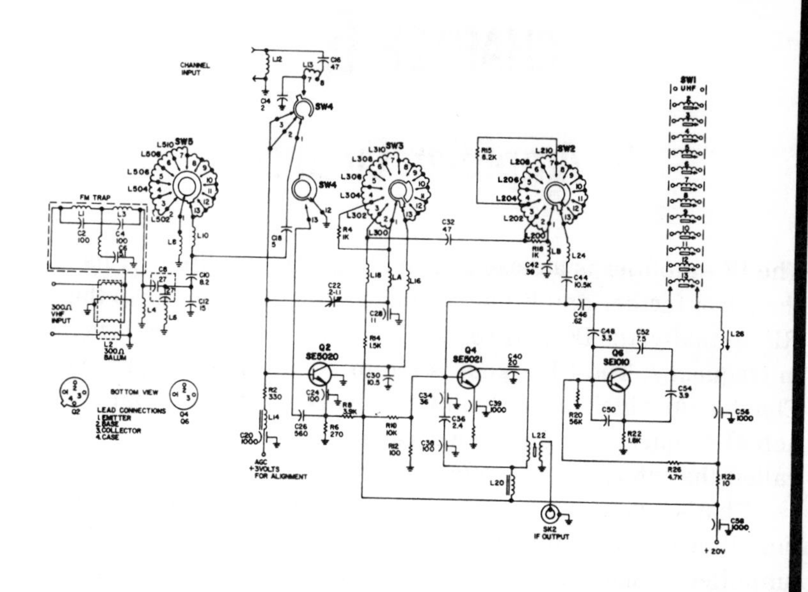

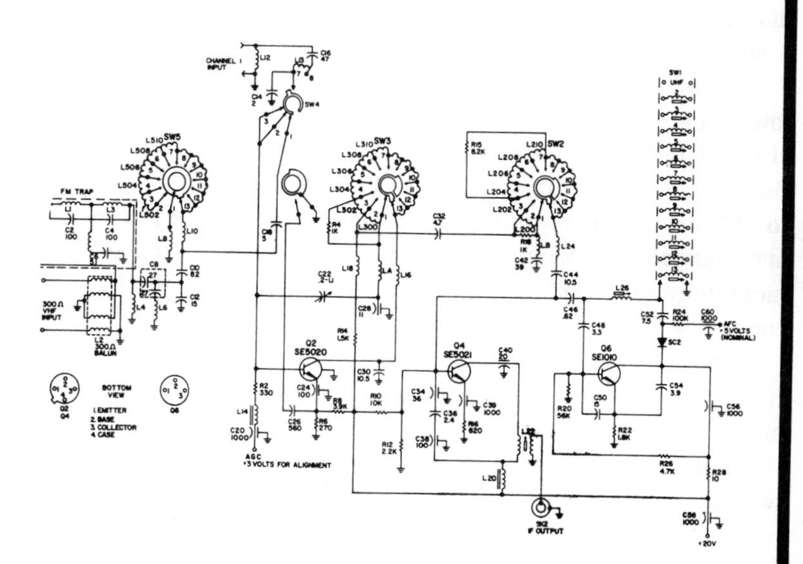

Fig. 5-11. Sylvania tuner schematic and alignment procedure.

CHAPTER 6

IF Amplifier Circuits

The IF amplifier is a close cousin of the RF amplifier. It has the job of further amplifying a range of signals similar to the RF. Actually, the IF is still an RF signal; it's just a little lower in frequency. The RF signal is fed to a converter stage (see Chapter 10) which changes all RF signals, no matter what the actual frequency in the tuning range, to a single frequency called the intermediate frequency.

That's where the big advantage comes in. While the RF amplifier has to be tuned to a number of frequencies, the IF amplifier is tuned to only one. The RF amplifier has the tricky job of trying to amplify every single station, up and down the dial, in a similar way, but the IF amplifier only has to amplify a single intermediate frequency.

In an ordinary AM radio, all incoming signals, from the low end of the dial at 550 kHz to the high end at 1600 kHz, are changed to 455 kHz. In an FM radio, all incoming signals from the low end at 88 MHz to the high end at 108 MHz are changed to 10.7 MHz. In a commercial TV, all channels from two to 82 are changed to 44 MHz. In any receiver that uses an IF, all the incoming signals are changed to the single IF frequency by the converter stage.

This conversion gives the IF amplifier a large advantage. It can employ selective gain, thereby providing linear amplification over the entire incoming bandpass. The number of IF amplifier stages used in any receiver may vary from one, two, three, or more, depending upon the application of the receiver.

IFs are designed to do a three-fold job. A bandwidth sufficient to pass the entire channel is required. Secondly, even though they pass the tuned-in channel, the IF stages are able to reject, decisively, any and all channels adjacent to the tuned channel. Third, the IF usually contains trap circuits that are intended to pick out a specific frequency in the transmitted channel and reject it. For instance, a TV IF has a sound trap to get rid of a large portion of the transmitted audio so it can't interfere with the CRT display.

The IF amplifier actually provides the largest total gain in the receiver. An ordinary transistorized 3-stage IF is capable of providing as much gain as the tuner and output sections combined. But when a strong signal is coming in, the AGC shuts down a large portion of the IF gain. In fact, the AGC can reduce the gain almost to zero. At that time the gain of the tuner and the output stages are enough to produce satisfactory signals. However, when a very weak signal is tuned in, the IF stages open up as the AGC dictates and all the gain is used. The IF stages can then produce as much if not more gain than all the other circuits put together.

BANDPASS

When a large bandpass is required, such as for a 6-MHz TV signal or a 150-kHz FM signal, a number of IF stages is used not only for gain but so that each stage can pass and amplify a portion of the wide bandpass. This is called "stagger tuning." Each IF stage is tuned to a different section of the band. Of the 6-MHz TV band, for example, each stage is tuned to about 2 MHz of the 6 MHz. Each passes its allotted frequency segment and they are all added in the output of the IF. In an FM receiver, each IF stage is tuned to a third of the 150 MHz.

In an AM receiver, a single-stage IF can be used since the receiver has to pass only about a 10-kHz band. The IF selectivity need be only 10 kHz on either side of the 455 kHz IF. The stage passes all frequencies between 455 kHz and 465 kHz. In lots of small, portable radios with cheap, tiny speakers that pass only a small range of frequencies, the IF stage may have the ability to pass only plus or minus 2 kHz. A tiny IF transformer passes only the narrow range of frequencies between 453 and 457 kHz.

The IF stages must pass all the required frequencies in a linear manner. If it doesn't, reproduction of the output will be distorted. The requirements vary considerably. As mentioned, the AM radio needs to pass only the frequencies the speaker can reproduce. Similarly, FM receivers need to pass only the frequencies that will be heard. There is no reason to pass stereo in an FM receiver with one audio output stage. Also, there is no sense in a black-and-white TV having to pass the color subcarrier, since it is not going to be used anyway. The basic IF amplifier has to be only as good as the rest of the receiver.

TRANSISTOR IF STAGE

Three types of devices are used in an IF amplifier stage—tube, bipolar transistor and FET. They all amplify and the associated circuits are quite similar, except for one consideration. In a tube the input impedance is high. In a bipolar transistor the input impedance is low. In an FET the input is high like a tube. The development goes full circle.

IF TRANSFORMER

Coupling from the tuner and between the IF stages is provided by transformers. The tube and FET use stepup transformers for coupling to match the impedance. Transistors use a stepdown transformer (Fig. 6-1). IC IF stages use external stepdown transformers to hook one IC transistor stage to another.

Whether the transformer is stepup or stepdown, it has certain characteristics. It is tuned to a particular frequency for the coupling. The coils are given a particular value of Q. (Q is a ratio that determines the exact selectivity of the coil.) A high Q increases the selectivity; low Q broadens selectivity and a wider band of frequencies is passed (Fig. 6-2).

What exactly is Q? It's a figure determined by the ratio of the inductive reactance to the AC resistance. Typical Q figures range between 50 and 200. A 50 figure is a fairly flat response over a wide range of frequencies. A 200 Q narrows the range of frequencies considerably and raises the response voltage many times higher than the 50 figure can produce.

The Q is greatest at the tuned frequency of the IF transformer. As the frequency gets away from the tuned point, the

Q drops drastically. The graph of the condition shows a hump with side skirts, and the selectivity of the transformer is controlled by the slope of the skirts.

In each transformer there are two coils, each with its own Q. The two coils are coupled together, and between the two coils the signal is passed from stage to stage. According to the physical space between the coils, the coupling will vary and produce different types of signal transfers (Fig. 6-3). Therefore, the coupling must be exact in order for the correct bandpass to be fully transferred and amplified. The coupling can be changed by varying the distance between the two coils of the transformer.

Consider a graph that has a horizontal axis representing frequency and a vertical axis representing the secondary output voltage. The frequency range chosen can be anywhere in the spectrum. The same percentage of variation will occur no matter what the frequency.

Now, let's take a look at a curve produced by a very loose coupling (Fig. 6-3). Notice how broad it is; it covers a wide

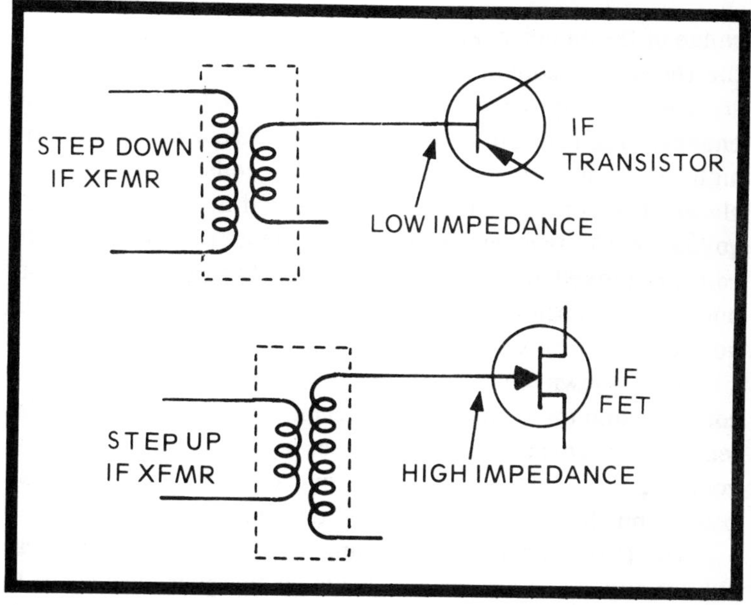

Fig. 6-1. A transistor has a low input impedance and needs a stepdown IF transformer input. An FET has a high input impedance and needs a stepup IF transformer input.

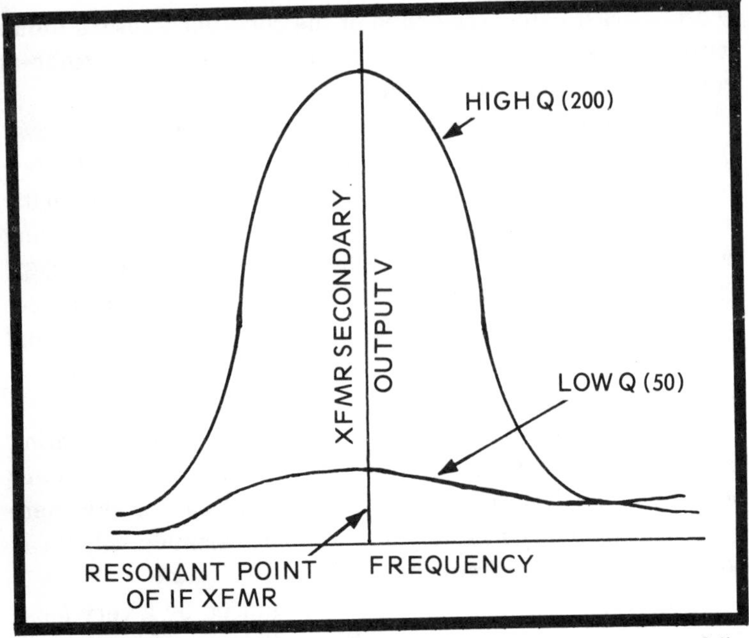

Fig. 6-2. The relative Q of a transformer determines the efficiency of the coupling of the resonant frequency.

range of frequencies, although there is a low-voltage peak at the resonant frequency.

Next, we move the coils together a bit and the curve changes. The range of frequencies narrows considerably and a high voltage peak is produced. Then, the coils are moved even closer. The curve broadens this time and instead of a single voltage peak, two peaks become evident. Lastly, when the coils are moved as close as possible, the two peaks spread out and become pronounced. At the resonant frequency the voltage dips almost to nothing.

It is shown that the passband increases with closer coupling, and the transformer's secondary output voltage also tends to increase; that is, up to a point called "critical coupling." At that spacing, the overcoupling produces two peaks and the secondary voltage decreases.

An IF transformer, according to its requirements, can exhibit any of the response characteristics just discussed. In an AM radio with one IF stage, the coupling is made to produce maximum voltage output with good skirt selectivity.

In a TV, degrees of overcoupling are used in order to pass the wide 6-MHz passband.

In a TV, the overcoupling, by itself, could hurt the response. So in order to get the wide response with overcoupling and not let the voltage drop to a minimum, the transformer is "loaded." All that means is that a resistor is placed in parallel with the secondary coil. This lowers the twin peaks a bit and elevates the drop in the curve at the resonant frequency.

AMPLIFIER

The coupling between stages is vital, but the actual amplification is performed by the amplifier device. In a single IF stage, the amplification can be as high as a 1000, but this type of amplifier can process only a very narrow band of frequencies. When a wider band is needed, more than one stage is used. In a TV there may be two, three or four IFs. Each stage multiplies the incoming signal by its amplification factor. If the factor is 20, then each stage has the ability to multiply by 20. If there are three stages, then 20 x 20 equals 400, and 20 x 400 equals 8000. That means a possible gain of 8000 can be produced. The AGC, however, limits that possibility. In actuality, a receiver never produces that great amount of

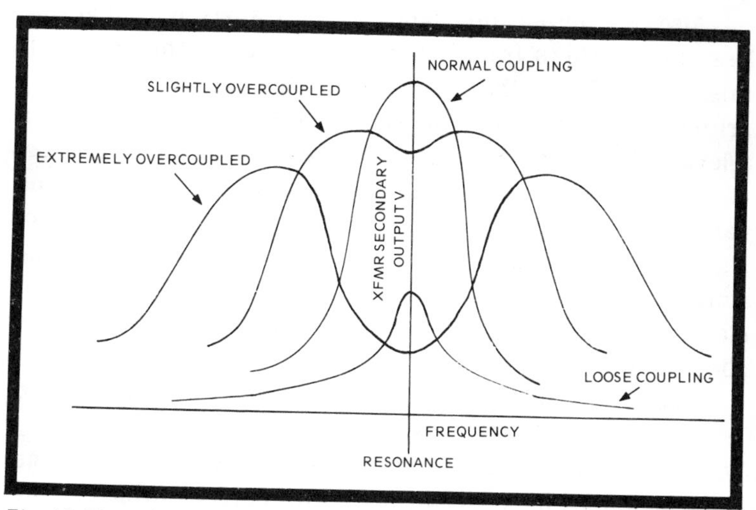

Fig. 6-3. The actual coupling distance between the transformer's primary and secondary determines the shape of the bandpass.

gain, since it is not necessary. The AGC sets up the actual amount of gain.

The typical IF amplifier stage in Fig. 6-4 is a common-emitter circuit, just as a tube IF uses a common-cathode setup. An emitter resistor is used to set up the bias. The resistor is bypassed so no signal can be developed across the resistor, just a DC potential. The base is attached through the stepdown secondary winding to the center of a voltage divider composed of a few resistors. The appropriate base voltage is obtained. The AGC is also hooked into the junction at the bottom of the secondary. Reverse bias is typical, although forward bias AGC is used, too, in some receivers. The potential difference between the emitter and collector is normally 4 volts, approximately. The PNP transistor is forward biased and a minute current flows when there is no AGC.

The stage's output transformer primary is designed to match the output of the transistor during the no-AGC condition.Best gain occurs during the time no AGC is fed to the base. There is little or no AGC during the time that there is no signal or an extremely weak signal. This allows the receiver to run at maximum during these times. As soon as some signal appears, the AGC begins operating. The reverse AGC cuts the gain drastically by lowering the collector current.

Also, the output impedance of the transistor changes as less current flows from collector to emitter. This causes a mismatch between the transistor and the output transformer primary. The mismatch helps reduce the gain quickly to a tolerable level. This mismatch also reduces the loading effect on the transformer which increases the voltage output and narrows the passband. With a strong station tuned in, plenty of signal still gets through. The narrowing of the passband simply increases selectivity and makes the desired signal clear. All adjacent-channel and spurious interference is eliminated.

That's why reverse AGC is preferred in a transistorized IF. When forward AGC is used, the opposite condition occurs. With an increase in signal, more forward AGC is applied to the IF. This makes the collector-to-emitter current stronger. The collector voltage drops with the bias increase and gain falls off. As the transistor-to-transformer mismatch occurs, it lowers the output impedance. This makes the IF passband wider. Adjacent-channel and spurious interference can enter the stage easier.

STABILIZATION

There is a substantial amount of distributed capacitance between the collector and base of a TV or FM IF amplifier, just as there is in the RF amplifier. The capacitance is a disturbing amount due to the high range of the frequencies being processed. In an AM radio working down in the kHz area the interelement capacitance usually can be ignored. In the high ranges, however, enough signal can be fed back from collector to base to start the stage oscillating. These parasitic oscillations can cause ringing in the sound and video. Whistles or ghost like images that vary during tuning are the result.

Neutralization must be added to the circuit (Fig. 6-5). A certain amount of signal is tapped off from the bottom of the output primary and fed back to the base through a small variable coupling capacitor. If oscillation takes place, the capacitor can be adjusted until the correct amount of the inverted signal, properly phased, cancels the feedback through the distributed capacitance.

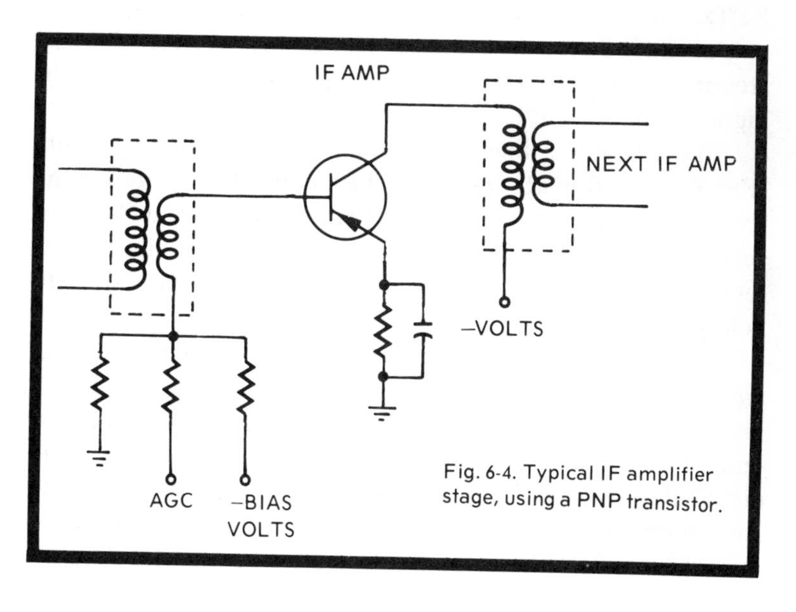

Fig. 6-4. Typical IF amplifier stage, using a PNP transistor.

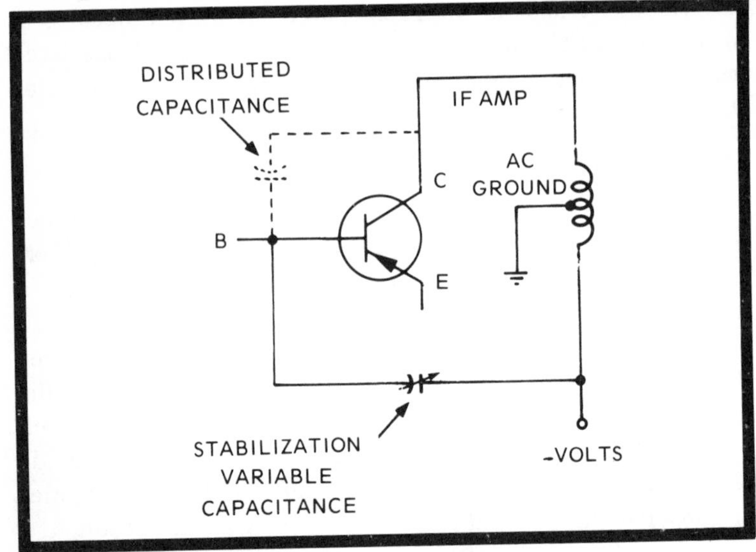

Fig. 6-5. An IF stage can easily start oscillating due to the CB feedback capacitance. Therefore, neutralization is needed.

COUPLING COMPONENTS

Transformers serve as the major coupling method between IF stages. The transformers used, as mentioned before, are stepdown types for bipolar transistors and stepup types for FETs. The stepdown is needed to match a high-impedance output into a low-impedance transistor. Conversely, the stepup is needed to match a lower impedance output into the high-impedance FET.

In the coupling there are a number of other components (Fig. 6-6). First of all, there may be a capacitor at a tap in the primary of the transformer. This capacitor is grounded and sets the AC level at the tap at ground potential while the DC level is maintained, thus providing the correct transistor DC voltages and polarities.

The capacitor grounds the signal at the juncture. That way, the two ends of the primary have inverted signals with respect to each other. Since some of the signal at the collector end of the primary is feeding back to the base through distributed capacitance, some of the inverted signal at the bottom end of the primary can now be fed back to the base, too. This cancels the unwanted collector-to-base feedback.

At various places between transistors are a number of capacitance-inductance tank traps. The traps typically have a tunable inductance. When the trap is tuned to its resonant frequency, it becomes an extremely low impedance in the coupling at the designed frequency. Any signal of that frequency will be sucked right out of the overall bandpass. Traps are tuned to adjacent-channel signals and do not let unwanted signals get through to the detector. The traps effectively increase the selectivity of the IF amplifier. The traps are typically connected from the collector to ground or the base to ground. Each trap forms a shunt for a particular frequency.

In a multiple-stage IF, it is possible that the last stage does not have transformer coupling. In such cases, a capacitor is used to couple the signal to the last stage. The last stage usually has a much lower impedance output load than the other IFs because the detector is a very low impedance. When the output load is so low, it is a characteristic of a bipolar transistor that its input impedance automatically becomes larger. It works out that a feasible capacitance voltage divider can be formed between the coupling capacitor and the input

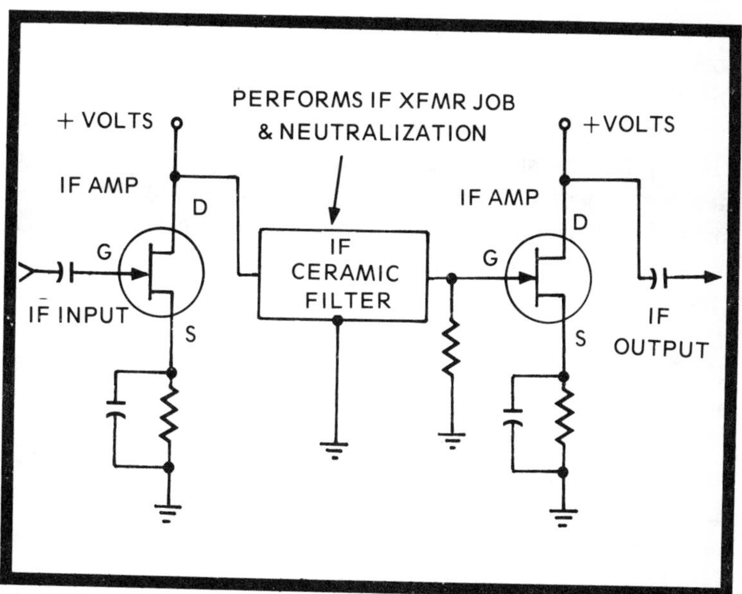

Fig. 6-6. A ceramic filter is sometimes used instead of a coupling transformer.

capacitance of the last IF transistor. The more convenient, non-tunable coupling can be used.

IF AGC

In tube-type IFs, AGC voltages are a tiny percentage of the multivolt potentials that are used to power the tubes. In transistor circuits, the AGC voltages are as much as some of the DC potentials at the E, B, C, source, gate and drain. Therefore, the AGC IF voltage is more of an influence in the IFs, so the requirements are stringent. There are AGC circuits that control all of the IF stages. Also, the control can be mixed between reverse and forward bias.

In a tube circuit, bias from an external supply can be substituted for the AGC, and if the external bias box cures the trouble, the AGC circuit probably contains the trouble. In a transistor circuit, such an easy technique is not always possible. Other techniques must be used. These are discussed in Chapter 12.

TESTING IF CIRCUITS

IF tests are similar to those recommended for other amplifiers as far as the DC potentials and bias voltages are concerned. That part of the testing, however, is usually not performed until the specific troublesome IF stage is localized. Localization is achieved with one of three techniques—signal injection, signal tracing or IF alignment.

Signal injection is best accomplished with a device like an EICO signal tracer for radios or a B & K Analyst for TV. The idea is to inject a signal into the IF input and observe what comes out. If you inject a similated IF signal into the IF input and a satisfactory signal emerges, either from a speaker or a CRT, the IF stages are good. If there had been no signal to start with, the trouble is localized to the stages previous to the IFs. Should no signal come out, though, the IF becomes suspect.

If the deflect appears to be in the IFs, inject the test signal, in turn, at the base or gate of the first IF, the collector or drain of the first IF, the B or G of the second, the C or D of the second, the B or G of the third and so on. As soon as the signal appears in the output, the trouble spot has just been passed over. Keep in mind that the further along the IF strip the injection is made, the weaker the possible output will be. This is because there is less and less amplification as each IF stage is passed over. Once the defective circuit area is isolated, DC voltage and resistance tests are begun to pinpoint the actual faulty component.

While signal injection is a valuable technique, it is somewhat risky, since the signal can be injected at the wrong spot or too much signal can be injected; either occurrence can cause damage to sensitive transistors. When using this technique, follow the instructions of the device's manufacturer carefully.

A more satisfactory localization technique is signal tracing. A signal has to be injected, but the injection site is not moved for each reading. Care can be taken to make the injection as safe as possible. The signal injected can be either a signal picked up by the antenna or the output of a sweep generator. The test signal is attached to the receiver antenna terminals. If you're using an off-the-air signal, tune for a strong local station or apply a few hundred microvolts from a generator to the antenna terminals.

Then an ordinary oscilloscope equipped with a demodulator probe is touched to the IF input. The gain of the scope has to be turned up at this test point since the signal is very weak. The demodulator probe, acting as an AM, FM or video detector, applies a detected output from the test point to the scope amplifiers. The sound or picture signal will appear on the scope, but only a tiny amount, if the tuner is good. If it's not, no signal will be displayed by the scope.

When a signal is present at the tuner output, the demodulator probe is touched on the B or G of the first IF, the C or D of the first IF and so on through the IFs. A good IF strip will produce a scope display at each test point. The display will show gain as each transistor output is reached. The gain will be considerable from the input of the first IF to the output of the last.

At the input of the first IF, a weak signal appears, as mentioned before. At the output of the first IF the signal should be quite large. At the input of the second IF, the signal will be the same amplitude as the output of the first. When the output of the second IF is tested, the signal should be considerably larger. At the input of the third IF, the scope should show the same amplitude as the output of the second IF. At the output of the third IF, the signal should show additional gain. And so on, throughout any additional IF stages.

Should the signal disappear at any of the test points, you have just passed over the bad circuit area. If a signal does not grow in amplitude, from the input to the output of an IF stage, that stage is not amplifying and there is trouble in the circuit. The signal will sometimes shrink from the input to the output of an IF stage. If that happens, the stage is defective.

ALIGNMENT

Aligning the IF stages is the most complicated procedure in any receiver (Fig. 6-7). Therefore, it is given a wide berth by most technicians and it is not uncommon to meet technicians who have never performed an IF alignment. It is a tedious, confusing, time-consuming procedure. Each manufacturer's service notes seem to be different. A technician will attempt an alignment, then discover that he does not have one of the required pieces of test equipment or jig attachments. Alignment, though, an invaluable technique, and the ability to intelligently perform an IF alignment, separates the top technician from the mediocre.

Alignment adjustments rarely change in a receiver under normal conditions. When it does change, it could mean a component has changed in value or has become defective altogether. A bad component can be pinpointed by attempting an alignment, and then when a transformer won't tune correctly or a coil won't produce a desired result, the troublesome capacitor is right near by. It could be that the inductance has opened or shorted. It could be a changed value resistor, capacitor or defective transistor. The alignment attempt will narrow the suspects down to a very few.

Factory notes (Fig. 6-8) are best for an actual alignment, but even though the notes appear to be different, they all follow a similar path. Basically, during alignment you drive a signal through the IF stages. The signal is amplified and then applied to the detector. Once the signal is detected, you can either listen to the signal from the speaker, or "look" across the detector load and observe what the signal is doing in peak form on a VTVM or in sweep form on the oscilloscope. Once the signal is detected, you do not need a demodulator probe on the scope. A direct or high-impedance probe will do the trick.

Radio

In an AM radio, an IF signal can be applied from a generator to the input of the converter stage in the tuner. The generator is set to the IF frequency, typically 455 kHz, and modulated by a tone, 400 Hz or so. The signal passes through the converter stage, through the IFs, and then is detected and sent on to the audio amplifier and the speaker.

Next, you turn the volume control to maximum and adjust the generator output until a comfortable tone is coming from the speaker. Then try the tuning capacitor. If the signal you are sending through the radio is truly at the IF, tuning the radio will have no effect on the tone from the speaker. If the tone is affected by the tuning, the test signal is heterodyned with the local oscillator. If the tuning dial does affect the tone, keep on adjusting the generator frequency until it doesn't. Be sure of this fact.

The actual alignment consists of adjusting the input and output IF transformers for maximum gain and clearest tone. Start with the last IF bottom coil, next the last IF top coil and so forth on to the first IF input transformer coils. If the alignment "takes," the tuned part of the stage is good. If the alignment seems to have no effect, there is trouble in the tuned part of the stage.

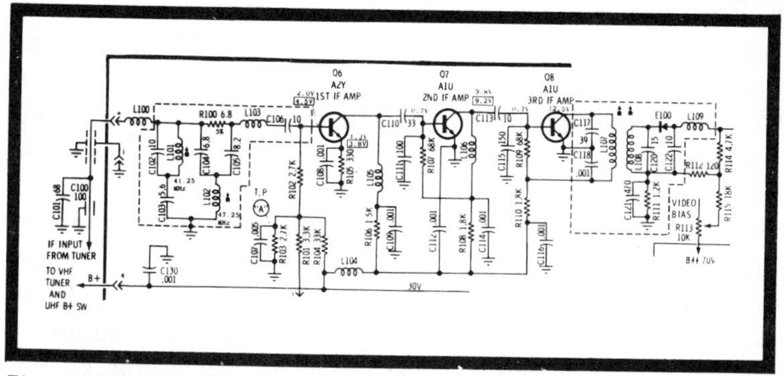

Fig. 6-7. Aligning the IF stages of a receiver is probably the most complicated job a technician is called upon to perform. (Courtesy Motorola)

Alignment of the TV IF strip must be performed according to the manufacturer's specifications as outlined in the notes. Both Sencore and B & K have developed generators that speed the alignment (Sencore SM158 and SM152 and the B & K 415, for example).

In transistor IFs, space is tiny, voltages are minute, devices are critical and, as a result, clues are hard to come by. DC resistance and voltages tests are almost meaningless. The only foolproof method to isolate subtle problems in the IF is with a sweep generator. It displays on the scope a picture of the signal that has passed through the IF stage. It shows the two parameters that the signal possesses—bandwidth and gain. Inaccuracies in bandwidth or gain indicates trouble in specific circuits. For instance, suppose the 41.25-MHz sound IF is not suppressed. Then the components that are supposed to do the suppression become suspect. They can be misaligned or defective. They are in the output of the third IF and are the sound reject pot and the 41.25-MHz sound trap. Try to align them. If they will align, the trouble has been clearly indicated and the cure made. Suppose the traps won't align. Check the shunt components. That's the way the alignment procedure can indicate trouble.

The big problem with TV alignment lies in the many actual service "moves" which necessitate all types of jig hookups and looking back and forth from TV to service notes. A good alignment from beginning to end can take a few hours. In a factory where the same TV is being worked on, day in, day out, a permanent alignment setup is made and the technician does the same job over and over. It becomes a fast procedure. In the independent shop the technician sees different sets constantly and the new physical appearances alone slow him down. What can be done?

Well, if you are in the business of occasionally having to align transistor TV IF strips, you must follow the factory way of doing things as closely as possible. Space should be provided in the shop specifically for alignment. The equipment could be put on a roller bench or cart so the service bench space can be utilized for other things, but a definite alignment setup must be made. The following general procedure is based on the similarities in factory alignment procedures. Of course, the manufacturers' recommendations should be followed to the letter.

Sweep Generator Connections

The sweep generator output signal is connected to the TV antenna terminals. However, the connection is not just clipped on like a VTVM. The generator is producing a signal like a transmitter and the connection must match the impedance of the receiver input. The usual generator output is applied through a shielded cable with an impedance of about 72 ohms. If the TV has an input of 72 ohms, attach it directly. Most TVs, though, have an impedance of 300 ohms. Therefore, a matching pad must be made to match the 72 ohms to the 300. This is the little pad of resistors that some manufacturers specify.

For the IF alignment by itself, the generator has to be attached to the converter input. The converter input impedance varies according to make and model. There is usually a test point available somewhere at a convenient spot on the tuner. The service notes will designate the location of the test point and also tell you what kind of resistance matching pad you need to match the generator to the converter. Some alignment procedures require you to attach directly into an IF stage. Again, the service notes will tell you where the connection should be made and what the matching resistance should be.

Scope Connections

The IF test signal is attached to the TV, processed through the IF circuit and then detected. The modulation is the sweep part of the generator output, along with marker frequencies that also modulate the IF. The video detector gets rid of the IF signal, and the sweep and marker signals are developed across the detector load resistor. The scope is attached to the video detector output. Here again, though, the scope is not just hooked across. An isolating resistor, typically 47K, is attached

PRE-ALIGNMENT INSTRUCTIONS

Before alignment of the video IF section is attempted, it is advisable to thoroughly check the system. If alignment is attempted on an IF section in which a faulty component exists, successful alignment will probably be impossible and the entire procedure will have to be repeated when the real cause of the trouble is corrected. Preliminary tests of the system should include voltage and resistance measurements, routine checks for bad soldering connections and visual inspection of the circuits for over heated components as well as for obvious wiring defects.

VIDEO IF & MIXER ALIGNMENT

Preliminary Steps

1. Maintain line voltage at 120 with variac.

2. Disable horizontal sweep by unplugging yoke leads.

3. Disable local oscillator by setting tuner between channels.

4. Apply the positive lead of a 6.2 volt bias supply to IF AGC (T.P. "A") buss and negative lead to chassis ground.

5. Check for correct 1st video amplifier bias by measuring 2nd video amplifier collector voltage. Voltage should read 20V with no signal input. If necessary, adjust bias by bypassing the 3rd IF collector to ground thru a .001 mf capacitor and adjusting the video bias control for 20V on the 2nd video amplifier collector.

6. Set the contrast and brightness control at maximum (extreme clockwise position). Set optimizer control maximum clockwise.

7. Short across tuner input terminals.

8. Maintain 1 volt peak to peak at the base of video amplifier except when specific values are given in the procedure chart

9. Refer to "Video IF and Sound ^lignment" detail for component and test point locations.

NOTE: To reduce the possiblity of inter-action between the two tuning cores in a transformer or coil, each core should be adjusted for optimum response in the tuning position nearest its respective end of the coil form.

SOUND ALIGNMENT
(Station Signal Method)

Reduce signal input into receiver by disconnecting one side or both antenna leads from receiver. Signal should be reduced considerably until some background noise is present.

1. Adjust both cores of 4.5MHz trap and A.T.O. transformer T300 for maximum audio.

2. Adjust primary (top core) of FM detector transformer T301 for maximum audio.

3. Adjust secondary (bottom core) of T301 for best sound with least noise.

4. Repeat Steps 2 and 3 until no further inprovement is noted in the sound.

VIDEO IF AND MIXER ALIGNMENT PROCEDURE

STEP	SWEEP GENERATOR AND MARKER	INDICATOR	ADJUST	ADJUST FOR AND/OR REMARKS
1.	To 1st IF base thru 120 ohm resistor in series with .001 mf capacitor. Set sweep to 44MHz, markers as required. Short junction of L103 & C106 to ground. NOTE: Generator must be properly terminated	Scope to base of 1st video amplifier thru 47K ohm resistor	Both cores of 3rd IF transformer (L107 & L108)	Adjust for maximum response at 44MHz (see curve No. 1). NOTE: The 3rd IF transformer consists of two individual coils inductively coupled.
2.	To mixer T.P. M thru 120 ohm resistor .001 mf capacitor. Remove short from junction of L103 and C106 to ground	Same as Step No. 1	41.25MHz trap L101 47.25MHz trap L102	Minimum response at proper trap frequency. See curve No. 2. NOTE: Temporary reduction of bias and increase of generator output maybe required to see trap clearly.
3.	Same as Step No. 2	Same as Step No. 1	Mixer output coil L26 on tuner	Adjust for symmetrical response. See curve No. 3.

in series with the scope, so the input loading of the scope won't change the shape of the curve that is developed in the detector circuit.

The sweep generator sweeps from about 39 MHz to 48 MHz, about a 9-MHz range. The sweep rate of most generators is 60 times a second. Actually any, sweep rate could have been picked, but 60 Hz is a convenient frequency since it matches the usual AC power line frequency. The oscilloscope is set at the horizontal rate of 60 Hz. This can be obtained from the

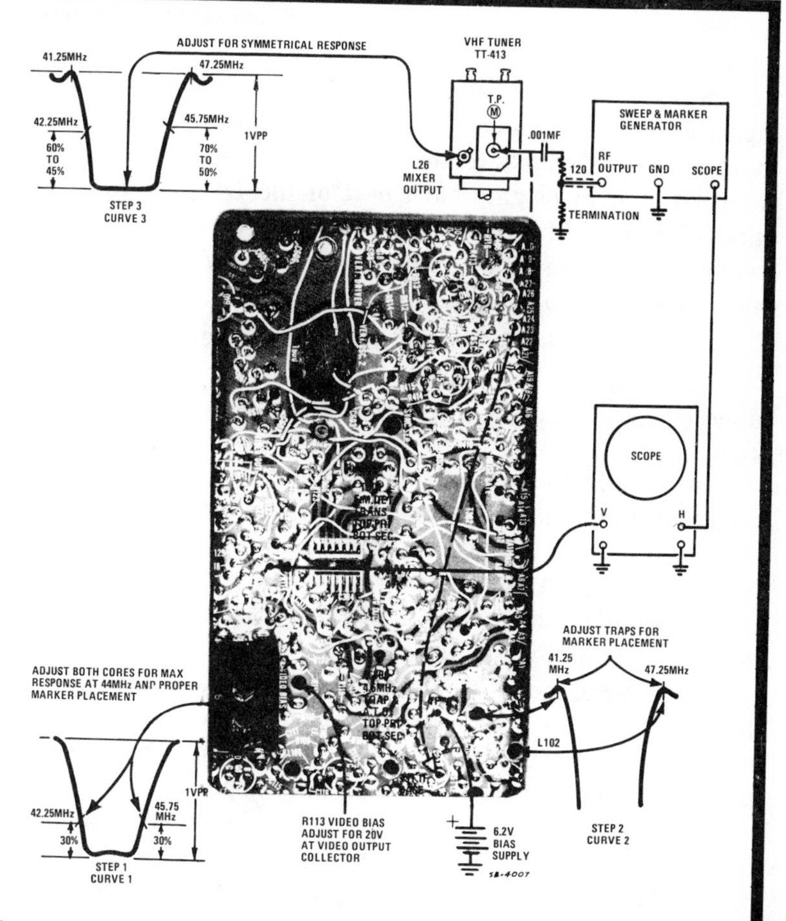

Fig. 6-8. Factory notes are best for an actual alignment since they were written specifically for the particular chassis.

generator itself or from the internal sweep in the scope. The 9-MHz sweep width is about what the IFs should pass. If it doesn't pass that much, or if a section of it is sucked out, there is a problem in the part of the circuit that is supposed to pass that range of frequencies.

The generator should be able to produce a "flat" range of about 15 MHz—that is, from about 36 MHz to 51 MHz. That way, as you are analyzing the bandwidth, if the IFs are misaligned, with a wide range of frequencies on either side of the required bandwidth. Defects or misadjustments will show up as a hump.

The test signal is applied to the oscilloscope's vertical input as the sweep frequency is sent in through the horizontal input. The signal causes vertical sweep on the scope according to to the amount of gain each part of the IF gives to each particular range of frequencies. Typically, the various signal frequencies are amplified or attenuated as they pass through the IF, according to this general plan (Fig. 6-9).

Starting from the high to low frequencies, they are:

1. 47.25 MHz, the adjacent-channel sound carrier. It is supposed to be trapped out so that it appears without amplitude or on the zero line.

2. 45.75 MHz, the video signal. Notice that it is only 1.5 MHz away from the adjacent-channel sound. That's why the sound must be correctly trapped or else it will be amplified. The video should appear about halfway up on the curve.

3. 44.25 MHz, the middle of the bandpass approximating the highest amplitude of the response curve.

4. 42.17 MHz, the color subcarrier. Notice that it is only 920 kHz away from the sound carrier (next in the listing). The color subcarrier should appear equal to the video but on the other side of the curve, halfway up on the slope. This signal is fairly strong and can heterodyne with the nearby sound carrier, producing a 920-kHz beat that can be detected, pass through the video amplifiers and appear as interference in the picture.

5. 41.25 MHz, the sound carrier, 920 kHz away from the color subcarrier. The sound carrier is exactly 4.5 MHz away from the video at 45.75 MHz. The sound and video carriers heterodyne, producing a 4.5-MHz sound intercarrier IF. A transformer tuned to 4.5 MHz picks up the beat in the second IF stage and sends it to the audio circuits.

The last IF stage is not involved with amplifying the audio. The last IF stage deals with suppressing the audio so it can't heterodyne with the color and produce the 920-kHz interference. The sound carrier, therefore, has to be suppressed in the response curve. Therefore, the 41.25-MHz sound is placed just above the zero gain line on the response curve.

6. 39.75 MHz, the frequency of the adjacent-channel video. This must be suppressed as much as possible; ideally, it should be on the zero base line of the curve.

IF Component Locations

In the IF strip, traps and transformers are strategically placed so they alternate or transfer, unaffected, the various signals so that the overall IF response approximates the preceding frequency conditions. In the input of the first IF is usually found the 39.75-MHz adjacent-channel video trap and the 47.25-MHz adjacent-channel sound trap.

The coupling transformer from the first to second IF is usually tuned to pass unaffected the 45.75-MHz video carrier. The coupling transformer from the second to third IF is usually tuned to pass the 42.17-MHz color subcarrier. The coupling transformer from the last IF to the video detector is usually tuned to the middle of the bandpass or 44.25 MHz. In the output circuit of the last IF is usually found a sound reject pot and a trap tuned to 41.25 MHz, the sound carrier.

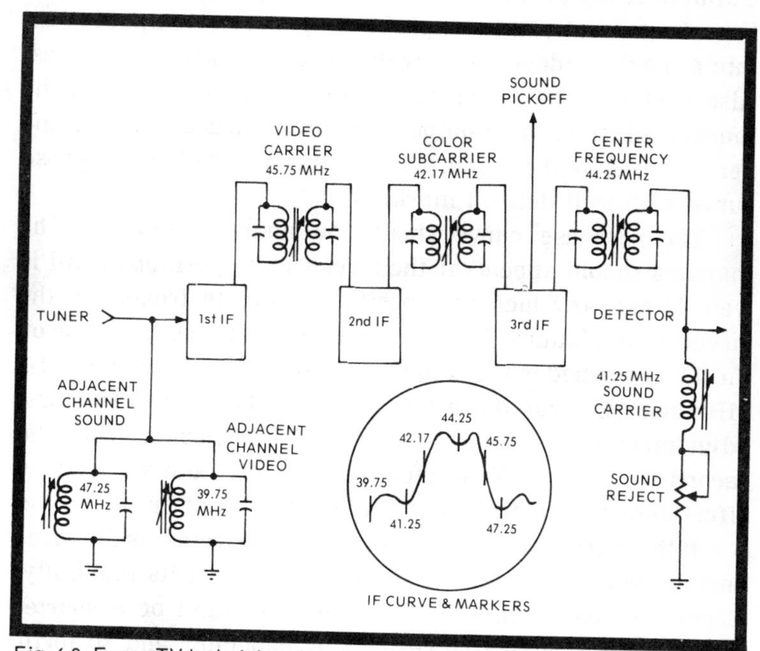

Fig. 6-9. Every TV technician should memorize these six adjustables and what they do to the IF curve.

MARKER INJECTION

In addition to the sweep generator and scope applications, the response curve needs to be "marked" at the frequencies that are being processed. Crystal-controlled oscillators that operate at a set frequency with no sweep are turned on and injected into the response curve. The marker signals can either be sent into the TV set and processed (preinjection) with the rest of the generator signal, or the markers can be added to the response curve after the sweep signal comes out of the TV on its way to the scope (post injection).

Preinjection of the markers can produce misleading results since the marker frequencies are processed in the TV. A trap could take out a marker or the marker could overload the circuits and destroy the response curve. Also, only one marker at a time can be preinjected; otherwise, curve can go all to pieces.

It's best to use a marker that adds the marking frequency after sweep signal has passed through the receiver. A good example of the post-injection marker method is found in the Sencore generators. A sample of the generator output is sent into a marker adder along with all the marker frequencies. Also sent into the adder is the TV response curve. In the marker adder the two outputs are combined and fed to the vertical input of the scope. The result is an excellent response curve with well defined markers.

The response curve should be analyzed closely. The markers should appear at their respective positions if all is well. If not, any incorrect positions indicate trouble in the circuit or a mistuned coil. For instance, suppose the color on the TV is washed out and the response curve shows the 42.17-MHz color marker down near the zero base line. The first adjustment to check is the coupling transformer from the second to third IF. Tune the cores slightly both ways in an effort to move the marker to the halfway point on the curve.

If the marker won't move and the curve remains the same during the adjustment, the second IF has lost its tunability. Check the components in the circuit. It might be a shorted transformer, a defective resistor or capacitor and so forth. The alignment has clearly isolated the circuit with the defect.

Local Oscillator

If the local oscillator is running during the alignment, it can introduce an interfering signal into the tuned circuits. It is a good idea to disable the oscillator, but it is tricky. The service notes usually remind you of the necessity of disabling the oscillator and how to do it. Some manufacturers recommend that you set the channel selector between Channels 12 and 13. If they suggest it, fine; however, in some tuners the oscillator circuit has a loading effect on the IF response. If you do the alignment without the oscillator loading, the curve will distort when the oscillator gets back into action. The point is, follow the manufacturer's suggestion as to what to do with the local oscillator.

Bias

The IF strip normally operates under the control of an AGC bias in order to maintain a constant gain, regardless of signal input level. When the IF strip is aligned, it is necessary to place an AGC voltage on the IFs. This is probably the most annoying part of the alignment. The rest of the alignment connections are quite universal, but the AGC input can vary 50 volts either way in solid-state TVs. Tube-type receivers, where AGC is negative, require only a volt or two. The hardest part is finding the location to attach the bias box. Usually, it's on the AGC side of the grid leak resistor.

With a solid-state TV it is a must to have the schematic so you figure out where the AGC is injected into the IFs. (See Chapter 12.) Once you find the right circuit connection, the question is, how much bias voltage, positive or negative? If you don't know the correct voltage, follow this procedure: Connect the bias supply and inject a positive voltage. Get some sort of a response curve on the scope. It will probably be way out of shape. Keep increasing the positive voltage slowly and watch the response curve. It will balloon out, but then start losing amplitude. Then it will start to assume its normal shape. When it looks normal, you have applied about the right amount of bias. While you are aligning, during the first few moves, adjust the bias a bit. You'll soon have it right on the nose. Even if it is slightly off, the alignment should be accurate.

CHAPTER 7
Audio & Video Amplifiers

The primary purpose of an audio amplifier is to send enough power to the speaker to make the cone move with enough force to reproduce the sound signal. The power is measured in watts. The primary purpose of a video amplifier is to produce about a 100-volt peak-to-peak video signal to drive the electron gun in a picture tube.

An audio amplifier has to handle only frequencies in the audio range, between 30 and 20,000 Hz. A video amplifier has to handle frequencies that produce a CRT display. In order to cut off the electron stream in an instant, so the display shows a picture element going from white to black, one four millionth of a second is involved. This means a frequency change of 4 MHz. Therefore, the video circuits process signals from about 30 Hz to 4 MHz.

However, despite these basic differences, audio and video amplifiers are serviced in a similar manner and the circuitry is very similar. The main difference between them lies in a few components in the video circuits to broaden the frequency response from a high of 20 kHz in audio to 4 MHz in video.

AUDIO AMPLIFIERS

Audio amplifiers have been around for a long time and are the most straightforward of circuits. There is nothing tricky in them and they haven't changed much in years. Even the amplification device changes from tubes to transistors and FETs leaves the circuitry almost the same. Whether the receiver is AM, FM, TV or what have you, the following circuit principles apply. Even an IC audio amplifier stage contains the same circuit configuration, although they are inaccessible.

Typically, there is a detector load resistor. This resistor, more often than not, is the volume control. The entire detector output is developed across the volume control. The top of the control, looking at Fig. 7-1, is attached to the detector and the bottom to ground. The center tap moves up and down the resistor, picking off as much or as little of the signal as is needed. At the top of the control, there is maximum volume and at the bottom little or no volume. A blocking capacitor between the center tap of the volume control and the circuit input relieves any loading effect and prevents a change in the frequency response as the volume control is varied.

In transistor circuits, the volume control acts as a current divider (Fig. 7-1). In FETs the volume control acts as a voltage divider (Fig. 7-2). In a current divider, the volume control feeds the low input resistance of the transistor. The voltage divider feeds the high input resistance of the FET.

In the current divider, the volume control changes the EB current by acting as the base resistor. In the voltage divider, the control changes the amount of voltage fed to the gate just like tube bias. Just as a transistor is a current-controlled device, the volume control varies the control current, and since an FET is a voltage-controlled device (like a tube), the control varies the bias voltage.

CLASS A AND CLASS B AMPLIFIERS

Most audio amplifiers operate at Class A or Class B. There is considerable confusion as to the classes. In transistor and

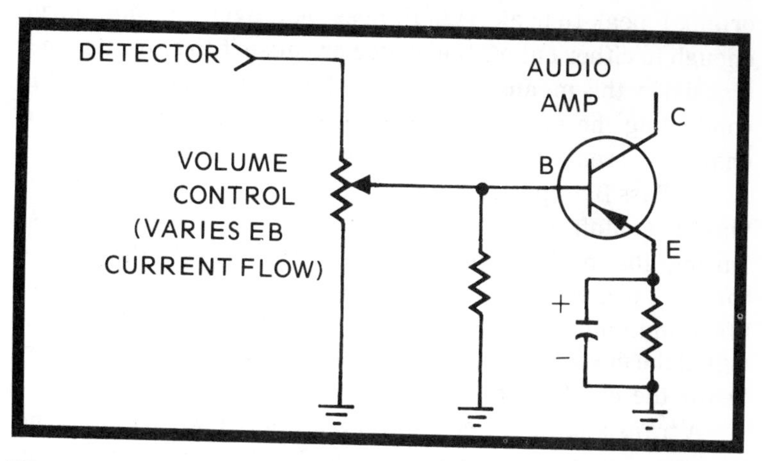

Fig. 7-1. In a transistor audio amplifier, the volume control acts as a current divider to the EB current flow.

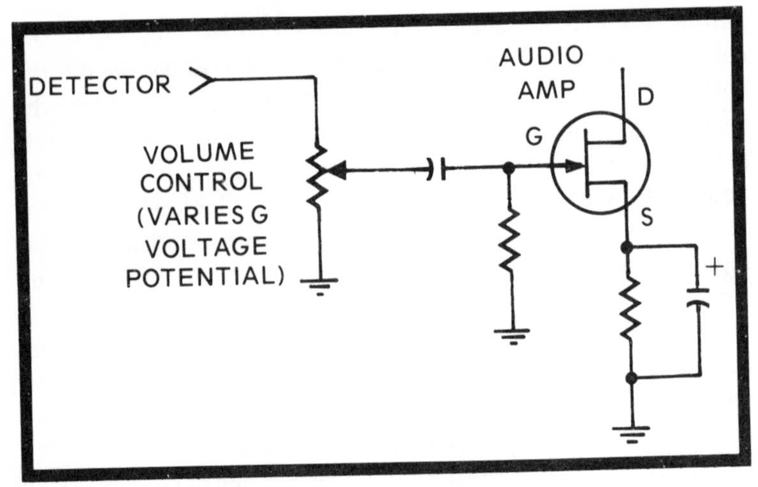

Fig. 7-2. In an FET amplifier, the volume control acts as a voltage divider to detector output.

FET Class A and B amplifiers, certain things happen, due to the fact that transistors are current-controlled devices and FETs are voltage-controlled.

A Class A amplifier is defined as an amplifier that is biased in such a manner that the input signal operates only over the straight portion of the device's characteristic curve (Fig. 7-3). This means that whatever signal goes into the input will come out of the output, amplified but not distorted. The original peak-to-peak signal does not ever become large enough to either cut off the device or cause it to saturate. In a transistor, this means the device is forward biased so that it is conducting the entire time the receiver is on. In an FET, a channel current flows at all times when the receiver is on.

A Class B amplifier is defined as being an amplifier that is biased at cutoff. As the signal is applied, here's what happens: During the positive part of the input signal, the device conducts and the positive signal appears in the output, amplified and undistorted. During the negative part of the input signal the device turns off, since this part of the signal appears below the cutoff point.

A transistor becomes a Class B amplifier when there is no DC bias. That way the positive signal puts a forward bias on the EB junction and the negative signal places a reverse bias on the EB. (Or vice versa according to whether the transistor is NPN or PNP; it doesn't really matter.) In an FET, the channel is placed at the pinch-off value, like a tube is placed at cutoff, and the signal turns on the channel during the positive swing and off during the negative swing. (Or vice versa according to whether the FET is an N-channel or P-channel type. It doesn't really matter which, the same principle applies.)

Class A amplifiers are always conducting. Class B amplifiers are cut off except when signal is applied. This can be a major design consideration for battery-operated audio amplifiers such as transceivers or walkie talkies. The battery will run down quicker in Class A circuits. In typical audio amplifiers it is common to employ Class B amplifiers due to this efficiency. When the signal applied is small, as in most applications, the current taken from the battery is small. Class A draws the same current whether the signal is tiny or great. As long as the unit is on, Class A circuits draw current.

VOLTAGE AMPLIFIER

In the typical audio amplifier, the detector output has to be amplified. The peak-to-peak signal voltage has to be amp-

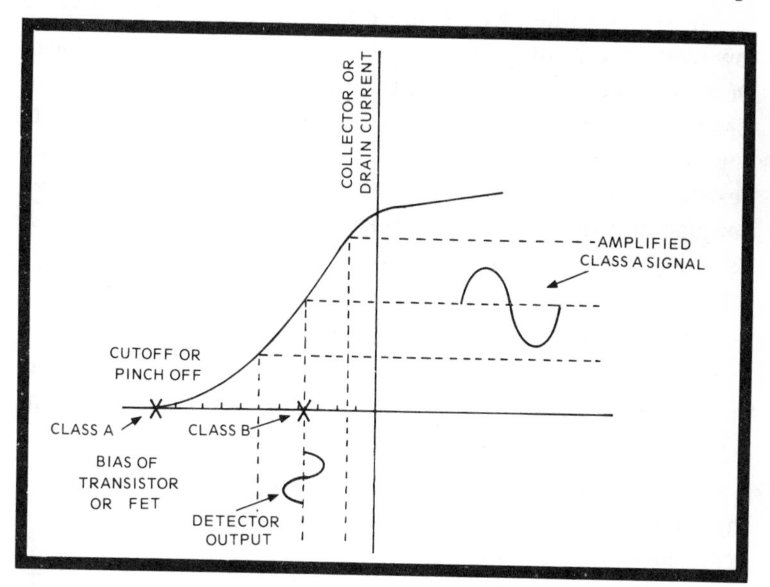

Fig. 7-3. In a Class A amplifier, all the peak-to-peak signal is amplified uniformly in the device.

lified between 50 and 100 times. The gain of the stage should be in that range. The volume control is in the base or gate circuit and the actual configuration is usually the simplest type of amplifier circuit, typically with Class A bias.

The volume control feeds the audio through the blocking capacitor and a peak-to-peak AC voltage appears across the base-to-ground resistor. The transistor or FET then receives and amplifies the signal. The emitter resistor is bypassed with a large-value capacitor in the filter range. A smaller value capacitor will not bypass the low audio frequencies. Only a capacitor near 100 mfd will.

The output of the AF voltage amplifier is developed across a resistor if the stage feeds a straight power amplifier, as in small radios. In most applications, though, the AF output is fed to the primary of a coupling transformer. The secondary of the transformer becomes the input of a pushpull output stage. The AF amplifier can also be called a driver for the pushpull stage.

POWER AMPLIFIER

In small radios where price is a major consideration, a low wattage output stage can be used to keep the price down. Of course, such a stage can't produce too much volume. The single power amplifier must be biased at Class A in order to reproduce the entire range of audio that is detected. A load resistor in the AF stage develops the signal and it is then coupled into the power stage through a blocking capacitor. A base or gate resistor to ground develops the signal and feeds it into the conducting device operating at Class A.

The output of the stage is fed into the primary of an output transformer. The purpose of the transformer is to match the output impedance of the device into the very low input impedance of the voice coil in the speaker.

The voltage amplification in the power stage is not more than 5 or 10. This is necessary because a heavy collector or drain current must be used in order to develop enough wattage to move the voice coil. The straight power amplifier and pushpull types are discussed in greater detail in Chapter 8.

VIDEO AMPLIFIERS

As mentioned at the beginning of this Chapter, a video amplifier has the job of developing a voltage that can be fed to a cathode ray tube. You'll find these stages with various names such as video driver, video amplifier and video output While the driver and amplifier names suggest voltage amplification, the video output name is misleading. Unlike audio or sweep outputs, the video output does not provide any power. The picture tube control grid never should draw any current. It is always biased negatively.

In order to modulate the cathode ray with the video signal, only the voltage between the cathode and control grid need be affected. A high peak-to-peak voltage on the order of 100 volts is required (Fig. 7-4) but no power is needed. The video output, therefore, like its preceding stages, is a voltage amplifier, unlike the audio output, which has to provide enough power to physically move the speaker cone. The cathode ray has no inertia. It stops and starts instantly and is intensity modulated with a voltage potential.

Whether the circuit uses a tube, transistor or FET, the problems are similar. The frequencies that have to be processed start in the vicinity of 30 Hz and can extend as high as 4.5 MHz in a high grade receiver. Small receivers can get away with only a 2.5- or 3-MHz response, but the same problem exists. The wide range of low frequencies require special components and design considerations. Some of the gain must be sacrificed in order to pass the wide range.

Gain is at maximum when the input and output impedances of the transistor or FET are matched. The problem is that distributed capacitance in the device becomes an appreciable load at the higher frequencies. In fact, an uncompensated video amplifier loses so much response that a clear video display is impossible.

The distributed capacitance appears across the collector-to-base junction and across the base to emitter. The same type of unwanted capacitance appears across the drain to gate and gate to source in FETs. These capacitances must be neutralized or else the video amplifier will not pass all the

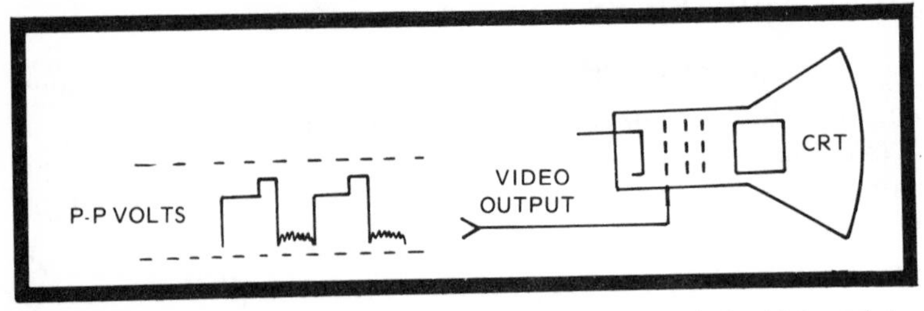

Fig. 7-4. The output of a video amplifier is not a current but a high peak-to-peak voltage.

frequencies. The following things are done to the video circuits to improve the response.

High-Frequency Compensation

First of all, the load resistors are lower in value. This immediately lowers the gain of the amplifying device and a wider range of frequencies is passed in a linear pattern. This, of course, is not enough, but it is a start. Next, a peaking coil is placed in series with the load resistor (Fig. 7-5). The peaking coil (a few turns of wire) is usually mounted on a resistor. The resistor eliminates the peaking coil's tendency to oscillate.

The peaking coil, though, is in shunt with the base-to-ground distributed capacitance in the next stage. The peaking coil and distributed capacitance are tuned to resonate near the high end of the video passband—around 4 MHz. This increases the coupled signal between the two stages; at this frequency this tends to flatten the response. Next, a peaking coil is placed in series with the coupling capacitor between the two stages. This coil and the coupling capacitor resonate near the high end of the range and further strengthen the response curve.

Lastly, a juggling of the coupling capacitor and coupling component values compensates for any low-frequency losses that might occur. The coupling capacitors are made a bit larger in capacitance than audio couplers, but the size is also determined by any bypass capacitors and other series load resistors that tend to prevent interaction between stages. Fig. 7-6 is a schematic of an FET video amplifier.

Other Jobs

The video amplifier has some other jobs to do in addition to driving the CRT. In many black-and-white TVs, the audio is not picked off until the video output. Then, a 4.5-MHz transformer receives the intercarrier sound, the 4.5 MHz beat between the video and audio signals. The sound pickoff transformer is usually right in the video output collector or drain circuit and has to be considered during video troubleshooting, even though it is a sound circuit.

Also in the video circuit, sync, blanking and AGC signals are taken off or inserted. These are also part of the overall video signal. See Chapters 12 and 14.

Another common component found in color TV video circuits is the delay line. The video or, as it's called, the Y signal, travels normally through a conventional video amplifier configuration. In a color TV, the color signals take another path through the color narrow band circuits. This slows them down in comparison to the wide bandpass present in the video circuits. The color and the video signals must arrive at their mixing station at exactly the same time.

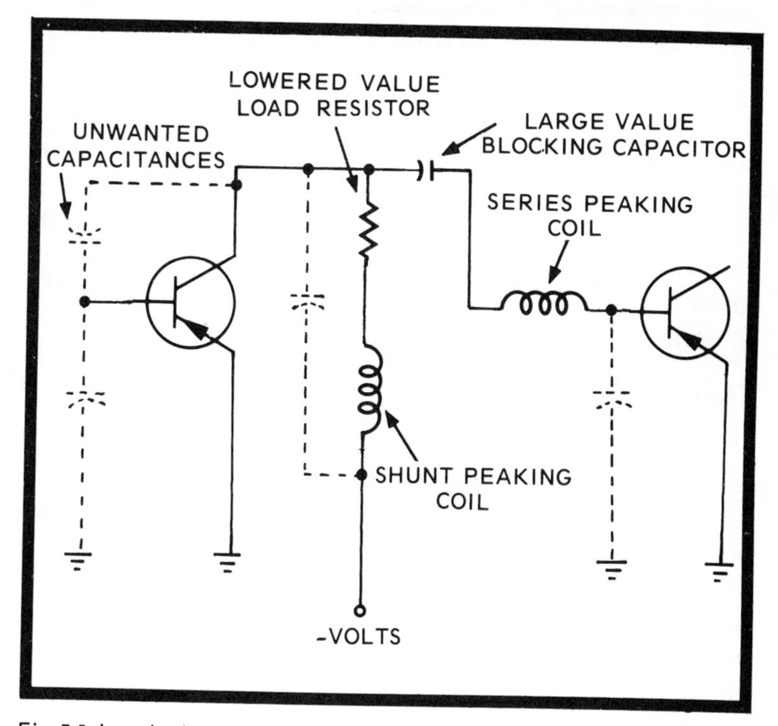

Fig. 7-5. In order to pass 4-MHz band, video amplifiers use peaking coils, low-value load resistors and large capacitors.

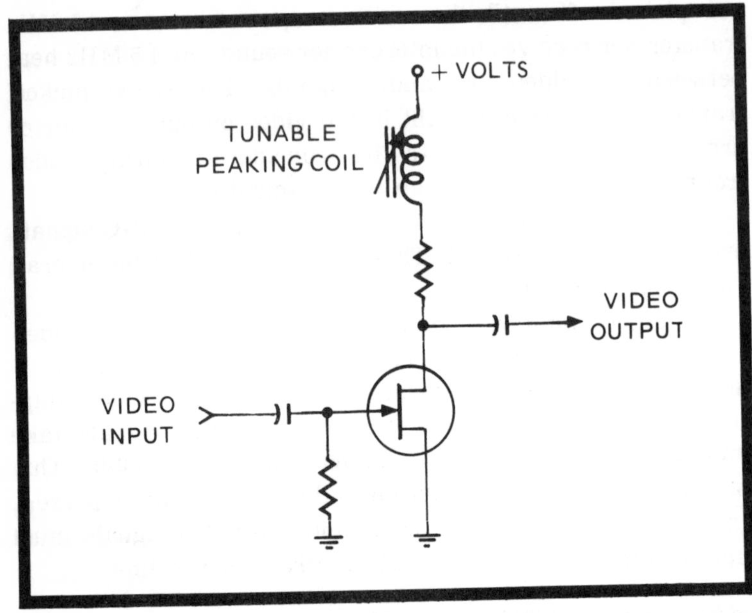

Fig. 7-6. An FET video amplifier can use a variable peaking coil to accentuate the high frequencies.

A delay line is placed in series with the video amplifiers to slow down the video so it arrives at the mixing station (matrix) at the same time as the color signals. The delay line is a long cylinder which is an inductive reactance to stop the video for a fraction of a microsecond (about 0-6 microseconds). It is used only in color TVs, but all color TVs must have a delay line in the Y amplifier.

Controls

In the video amplifier circuits, the brightness and contrast controls are usually found (Fig. 7-7). The brightness control is often in the actual picture tube input. Typically, it varies the bias between the picture tube cathode and control grid in a single-gun tube. As the bias between the two elements becomes more negative, the intensity of the cathode ray is reduced, lowering brightness. When the bias becomes negative enough, the cut off point of the tube is reached and the brightness is extinguished. Conversely, as the bias becomes more positive, the brightness increases.

In a 3-gun CRT, the brightness control is found in the base of one of the video amplifiers. The device will probably be a sharp cutoff type. As the brightness control is turned down, the video amplifier conducts less. The 3-gun CRT is usually cathode fed with video, and the less video that is sent into the cathode, the more positive the cathode becomes. This increases the cathode-to-control grid bias, reducing the brightness; as cut off in the CRT is reached, the raster is extinguished. Conversely, as the video is increased, the cathode becomes more negative, reducing the potential between the CRT elements. This increases the brightness. The cathode becomes more positive when less video is passing through the video amplifier.

The contrast control changes the amount of gain in the video amplifier. Typically, an unbypassed resistor in the emitter or source leg acts as the contrast control. When the control is turned all the way off, the full value of the resistor is in the emitter circuit. The video signal is developed across the control resistance, but it is opposite in polarity to the signal

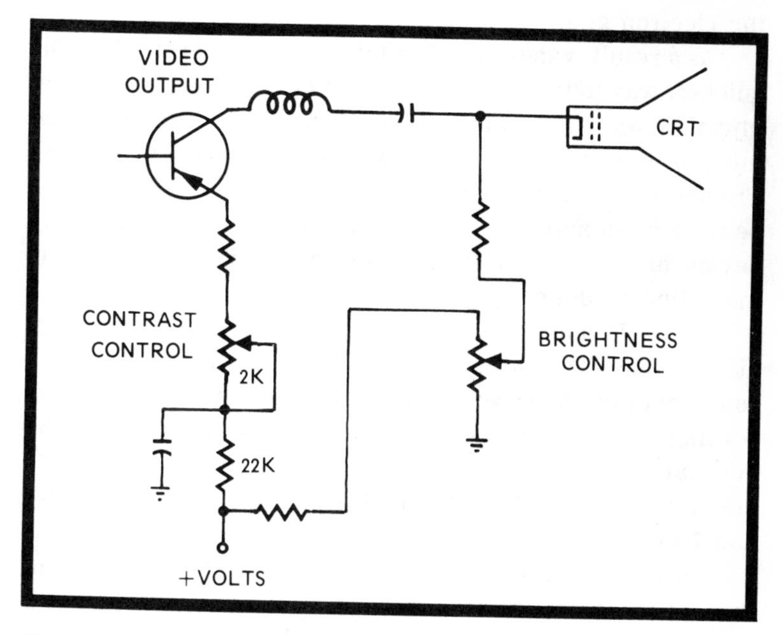

Fig. 7-7. The contrast control introduces degeneration, while the brightness control sets a DC bias.

being amplified. By degeneration, the emitter-developed signal cancels the original signal.

As the contrast control is turned on, the unbypassed resistor is gradually shorted out. The degeneration becomes less and less, and more of the video is permitted to be amplified. The total emitter resistance is about 25K. The contrast control is only about 2K. This small percentage of resistance in the contrast control, in comparison to the total emitter resistance, introduces the desired degenerative effect but does not appreciably affect the emitter DC bias voltage.

TESTING AUDIO AND VIDEO CIRCUITS

Aside from the frequency compensation components, audio and video amplifiers are quite similar. The volume control and contrast control act in a similar manner, each varying the gain of the stage. The brightness control is really a CRT bias control, and while part of the video circuit area, it is not in the video processing circuitry. It changes the DC level in the electron gun.

As a result, video amplifier tests follow general rules. The quickest way to localize the defective circuit area is by signal injection, although signal tracing can also be useful. Signal injection methods for audio and video amplifiers are identical. The same audio signal generator putting out a 400-Hz note can be used in an audio stage with the speaker as the indicating device or in a video stage with the CRT display as the indicating medium.

A 400-Hz note from a speaker, of course, is a pleasant audible tone. The same 400-Hz note on a CRT places black sound bars on the screen. The number of bars will vary according to the frequency of the note and the frequency of the vertical sweep rate. For instance, in a TV with the sweep rate at 60 Hz, a 400-Hz note will show 400 divided by 60 or between 6 and 7 black bars.

To inject the test signal, the first test point is the output of the last stage, whether it's the audio output or the video output. If the note is present on the indicating medium, the signal is passing through, and all of the circuit area through which the note is passing is good. If there is no indication, the signal is not getting through and there is trouble in that circuit.

The signal injection point is moved progressively forward toward the detector. As the input of an amplifying device, such as the base or gate, becomes a test point, the signal will be amplified. If the signal does not appear, or is not stronger, that stage contains the troublemaker.

For TVs there are special signal generators, like the B & K Analyst, that will put a test pattern on the CRT. In addition to just seeing something on the CRT, the familiar test pattern provides other video circuit information that helps pinpoint more subtle troubles than just a go no-go indication.

If the test pattern lacks contrast, a weakening of an amplifier is indicated. If there is a loss of high-frequency response, the vertical wedges in the test pattern will blur while the horizontal wedges still looks good. Some test patterns have the vertical wedges marked. The point where the lines blur is the limit of high-frequency response. In some cases, it is only as high as 1 MHz or 2 MHz.

If there is a loss of low-frequency response, the horizontal wedge will become less contrasty than the vertical wedge. Also, a loss of low-frequency response causes the picture to smear badly from left to right. These indications tell you what type of trouble is occurring in the video amplifier and the components that provide this type of activity are suspect.

Signal tracing can be exercised in the audio and video amplifier, using a scope as the indicating device. A test signal can be fed in at the antenna terminals and the scope used to trace its progress through the circuits. However, since these circuits end up directly at a speaker or CRT, the use of the scope is not necessary for the localization of trouble. Some technicians are expert on taking apart a scope picture and will insist on the signal tracing technique. All well and good. From an overall point of view, the speaker and CRT, using signal injection, saves a little time.

Once the defective stage is localized, individual components can be tested. Peaking coils can be shorted out momentarily (with the TV operating) if they are suspected to be open. If the picture improves, the coil is open. Peaking coils

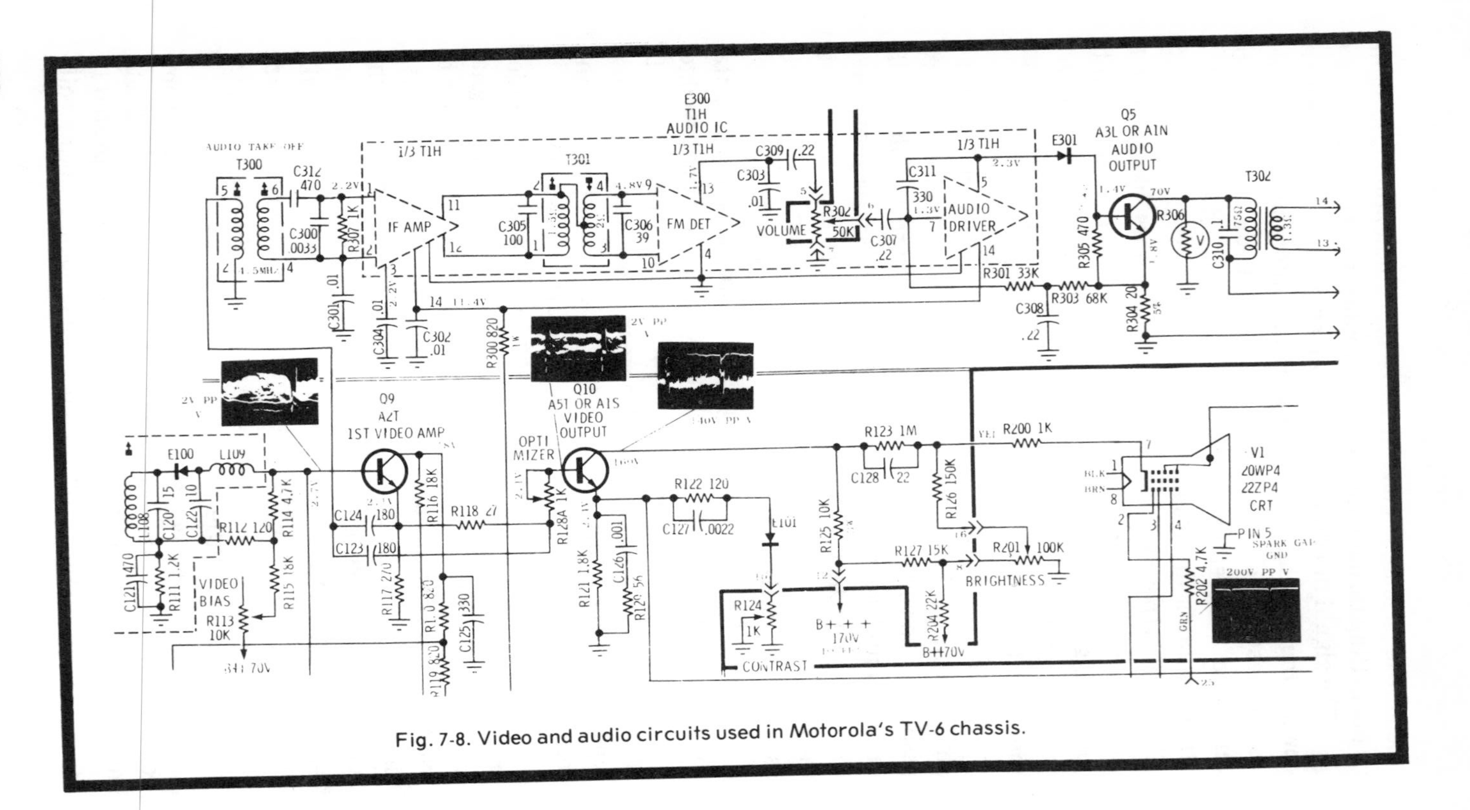

Fig. 7-8. Video and audio circuits used in Motorola's TV-6 chassis.

rarely open in any place other than at the connections. Most of the time a drop of solder on the open connection will fix it.

If the audio is blurry, or the TV picture is fuzzy, but a 400-Hz note on a test pattern injection produces a clear sounding note or a good pattern, the amplifier circuits are probably good and the trouble is an alignment defect in the IF or detector stage. Video detectors also have frequency-sensitive peaking coils and resistors and blocking capacitors that can change value. Fig. 7-8 is a schematic of the Motorola TV-6 video and audio circuits. Troubleshooting charts follow.

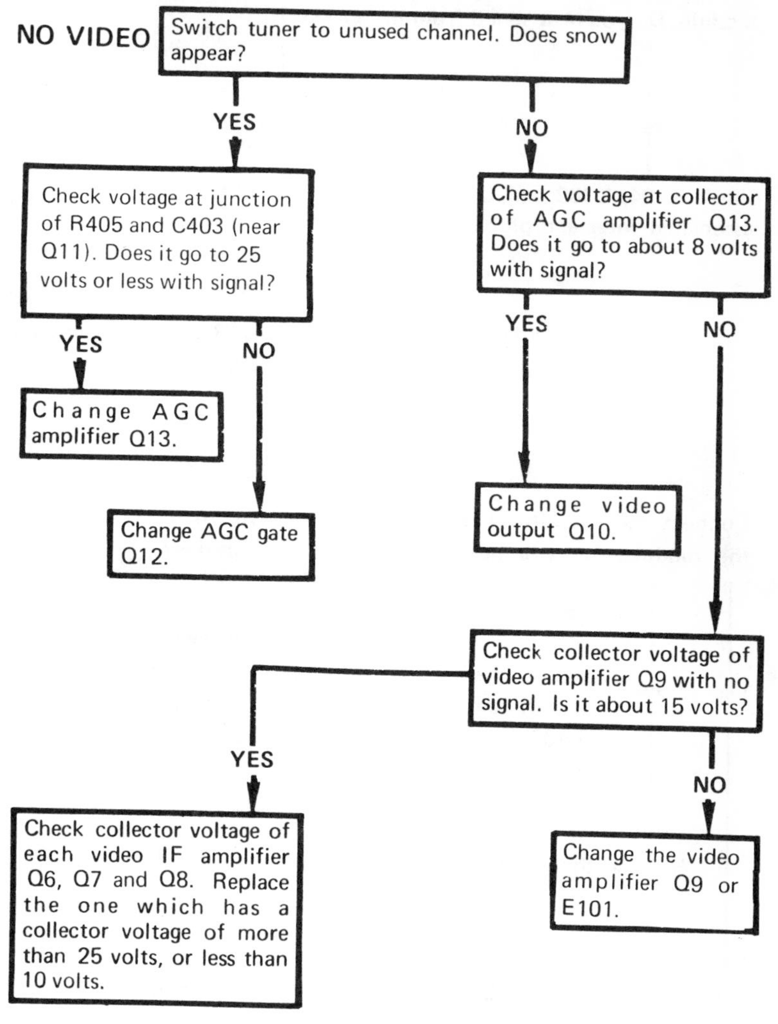

NO AUDIO - PICTURE OK

Note: To troubleshoot this section, use a 0.1uf/600v capacitor with test leads attached to each end.

Set the volume control to its midposition and connect the capacitor between the CRT, pin 8 and term. 2 of the audio module. Do you hear a hum from the speaker?

yes

Connect the capacitor between term. 2 and 8 of the module. Do you hear sound from the speaker?

yes — Check R304, C308 and the associated wires and plugs.

no — Check the voltage at terms. 3 & 4 of the module. Are the voltages about 1.6vdc?

yes — Connect the capacitor between TPIII and term. 4 of the module. Do you hear a buzz from the speaker?

no — If the voltage at term. 3 is OK, but the voltage at term. 4 is not, check L108 for an open circuit. If both voltages are off, remove the module and check for a short between L108 and chassis ground. If there is no short, replace the module.

yes — Check C123, L108, C128 and the module socket.

no — Replace the module.

Set the volume control to its midposition and connect the capacitor between the CRT, pin 8 and term. 2 of the audio module. Do you hear a hum from the speaker?

no

Connect the capacitor between the CRT, pin 8 and the base of Q301. Do you hear a hum from the speaker?

yes

Check the voltage at term. 6 of the module. It should be about 9vdc. If it is OK, replace the module. If it is not OK, check R301 and replace the module, if necessary.

no

Connect one end of the capacitor to chassis ground. Touch the other end to the collector of Q301. Do you see a spark at the point of contact and hear a pop from the speaker.

yes

Check Q301 and R306.

no

Check the speaker, earphone jack, and the secondary of T301 for continuity. Check from the collector of Q301 to +130 volts and chassis ground. The resistance between the collector and +130 volts should be about 250 ohms. The resistance between the collector and chassis ground should be 750 to 1500 ohms.

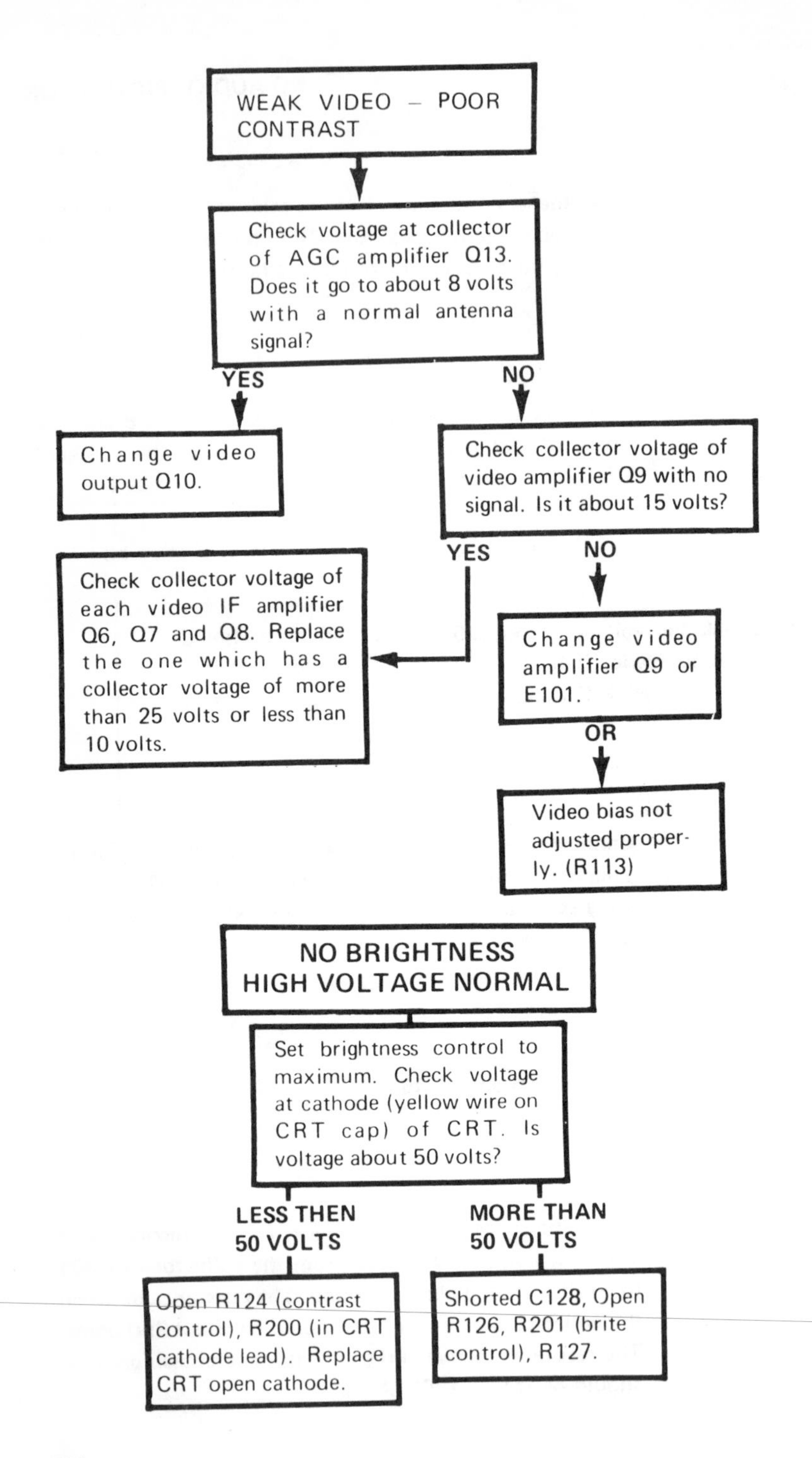
WEAK VIDEO – POOR CONTRAST
Check voltage at collector of AGC amplifier Q13. Does it go to about 8 volts with a normal antenna signal?
YES
NO
Change video output Q10.
Check collector voltage of video amplifier Q9 with no signal. Is it about 15 volts?
YES
NO
Check collector voltage of each video IF amplifier Q6, Q7 and Q8. Replace the one which has a collector voltage of more than 25 volts or less than 10 volts.
Change video amplifier Q9 or E101.
OR
Video bias not adjusted properly. (R113)
NO BRIGHTNESS HIGH VOLTAGE NORMAL
Set brightness control to maximum. Check voltage at cathode (yellow wire on CRT cap) of CRT. Is voltage about 50 volts?
LESS THEN 50 VOLTS
MORE THAN 50 VOLTS
Open R124 (contrast control), R200 (in CRT cathode lead). Replace CRT open cathode.
Shorted C128, Open R126, R201 (brite control), R127.

NO AUDIO – PICTURE GOOD

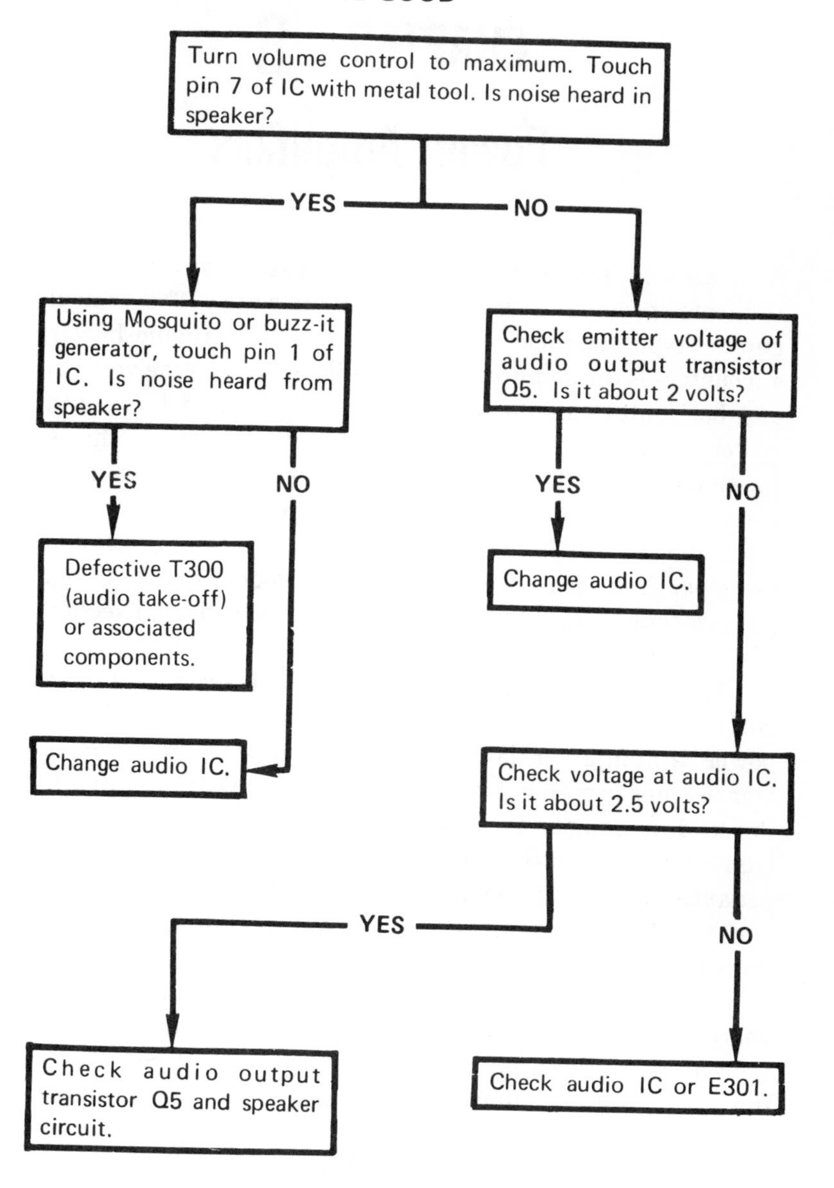

CHAPTER 8
Power Amplifiers

A transistor power amplifier does just what the name suggests—provides power measured in watts. The typical power type application in receivers a troubleshooter encounters is the audio output in a radio or TV and the vertical and horizontal outputs in a TV. In transmitters there is a power amplifier that drives the transmitting antenna. This is the end result of all the voltage amplification. All signals must eventually be put to work. For our purposes here, work means the actual expenditure of energy in watts. The only notable exception is a video output stage that produces only a high peak-to-peak voltage to drive the cathode-control grid section of a CRT's electron gun.

The output of a power amplifier is transformed to a current expenditure or a wattage consumption in a load. The power amplifier usually is coupled to the load via a transformer to match the impedance of the power amplifier to the load. A notable exception to this is the circuit where a transistor is attached directly to a load when the output impedance matches the load almost exactly. The transformer can be either stepup or stepdown type according to actual impedance demands.

The common way of converting voltage amplification to energy is by changing the AC voltage to an electromagnetic field. In an audio output stage the signal is developed across the primary of the output transformer, which is in the output leg of the device. The varying voltage induces another voltage in the secondary, which is attached to the voice coil. An electromagnetic field is developed in the transformer, and the size of the field is a direct result of the power amplifier's peak-to-peak output voltage. The higher the peak to peak, the larger the magnetic field becomes.

In a radio or the audio output of a TV set (Fig. 8-1), the output transformer secondary current is applied to the voice coil in the speaker. The voice coil is suspended within the field of a permanent magnet. As the output current flows through the voice coil, the field it develops interacts with the magnetic field and causes the voice coil to move in and out. In so doing, the voice coil moves the speaker cone to which it is attached, producing sound waves.

The vertical output stage in TV receives a sawtooth voltage from the oscillator. The amplifier is designed to operate at a bias, with appropriate waveshaping components, to form the sawtooth into the exact shape that is required for the vertical sweep. The shaped wave, with a particular peak-to-peak voltage, is fed into a vertical output transformer. The transformer has a prescribed number of windings to match the transistor power output into the vertical deflection coils.

The vertical deflection coils are wound in a yoke that is mounted around the neck of the CRT. The sawtooth current from the output transformer flows through the yoke coils and produces a magnetic field around the yoke. The size of the field varies in direct accordance to the amount of peak-to-peak voltage that is produced by the vertical output. Passing

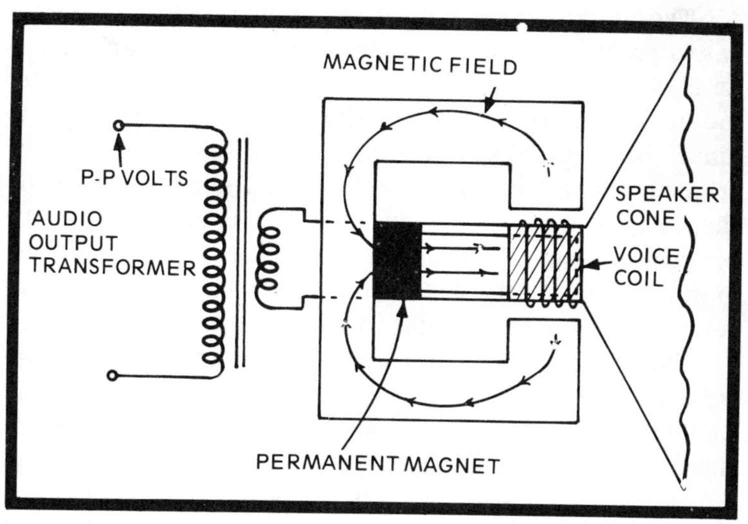

Fig. 8-1. An electromagnetic field moves the speaker cone, effectively transferring peak-to-peak voltage to energy.

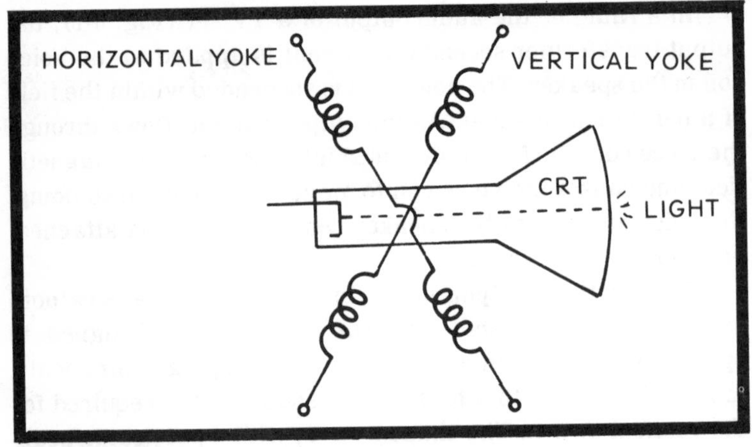

Fig. 8-2. The vertical and horizontal output stages drive the magnetic yoke field that moves the cathode ray in prescribed patterns.

through the influence of the yoke is the electron stream from the CRT gun structure (Fig. 8-2). The electromagnetic field around the yoke moves the electron beam in a vertical direction. Typically, the vertical yoke causes the electron beam to be drawn down slowly and back up quickly at a 60-Hz rate, thus the voltage amplification is transformed into the physical movement of the electron stream.

The horizontal output stage is a bit more complex, since it has to produce two outputs and one part of the output interferes with the other. The two outputs are the driving pulse for the horizontal section of the yoke and the driving pulse to make the high voltage. The driving pulse for the yoke is originated in a horizontal oscillator and is also a specially shaped type of sawtooth waveform that is amplified by the horizontal output transistor. The transistor output is matched to the horizontal coils of the deflection yoke by a horizontal output transformer, also called a flyback transformer.

The electromagnetic field formed around the yoke by the current from the output stage causes the electron beam in the CRT to be drawn slowly across the picture tube from left to right, and then retraced quickly to start the next line. The typical horizontal sweep rate is 15,750 Hz. The vertical and horizontal sweeps occur simultaneously, producing a full screen of light.

The flyback transformer also produces the high voltage that is applied to the anode of the CRT. The flyback pulse occurs during the retrace interval. The collapse of the magnetic field during the retrace produces a high DC potential in the flyback transformer. It is tapped off and rectified for the CRT anode.

PUSHPULL POWER AMPLIFIERS

In Chapter 7, single-ended power amplifiers were discussed. They are very similar to any audio amplifier operated in Class A. Another common way to produce power in an audio circuit is the pushpull power amplifier. It requires two transistor devices connected 180 degrees out of phase. As the signal goes positive, one of the devices turns on and amplifies. The other device is biased off. Then, as the signal goes through the zero line and becomes negative, the second of the devices turns on while the first one turns off. That way the signal is being processed at all times (Fig. 8-3). The current developed in the output is a result of the amount of signal in the input. Small amounts of signal turn the device on to a small degree, causing a small amount of conduction. Large amounts of signal cause heavy conduction.

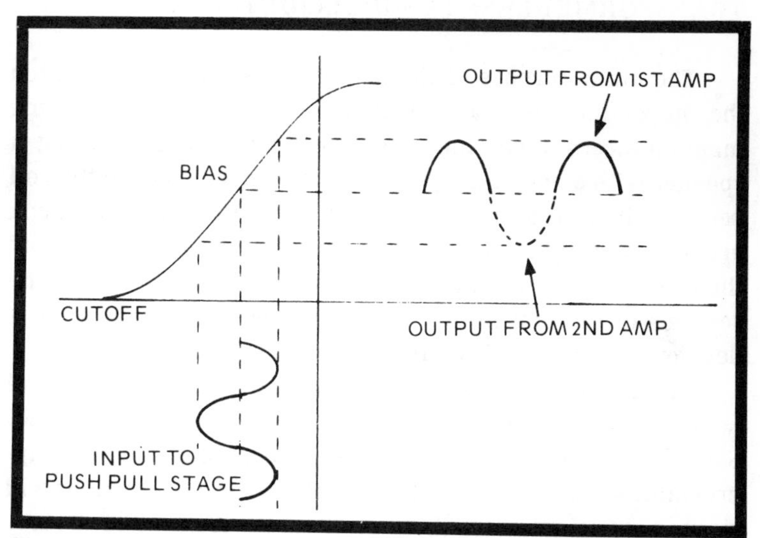

Fig. 8-3. Pushpull outputs are more efficient since the output transistors conduct only when amplifying a signal.

The input signal is transformer-fed to the pushpull stage. The primary of the transformer has a conventional layout. The secondary of the transformer, however, has a grounded center tap. This places both ends of the transformer winding exactly the same distance above and below ground. This permits the pushpull setup to inject the positive half of the signal in one transistor and the negative half of the signal in the other transistor. The transistor emitters are tied together and the bases are attached to the two ends of the input transformer.

There is also a transformer output. Its primary is center tapped, also making the signals at the two ends out of phase. One end of the secondary, though, is grounded. Therefore, as the top transistor conducts through the output transformer; the positive signal induces a current in the coil while the bottom transistor produces nothing. Then as the top transistor turns off and the bottom one turns on, a current is continuously induced. The output drives the voice coil with great efficiency during the full range of the signal. The transistors produce current gains since they vary the amount of current that is induced in the output windings.

TRANSFORMERLESS PUSHPULL OUTPUTS

The main purpose of the output transformer is to match the impedance of the transistors to the speaker. Some manufacturers have eliminated the transformer by using a speaker with a center-tapped voice coil, in which case the coil becomes its own output transformer. The two ends of the coil are attached to the collectors of the pushpull transistors and the center tap is attached to the power supply (Fig. 8-4). The sections of the coil match the transistor and the current is developed in the coil directly from the collector output.

The circuit works fine, but a servicer has to be careful never to turn on the unit while the speaker is not attached. Should the unit be on without the speaker load, and any appreciable signal is applied, the current in the PN junctions will rise to a sufficient amount to overshoot the transistors. Output transistors are the most expensive types of transistors, too!

VERTICAL OUTPUT CIRCUITS

The output type transistor works well in the vertical output circuit. Since it has a low output impedance in comparison to tubes, in a black-and-white receiver it is not uncommon to find a vertical output matching transformer is not used. The transistor is matched directly into the vertical deflection coils. In a color TV, this is not too convenient since vertical centering, convergence and pincushion circuits are fed from the vertical output. A complex matching transformer is needed for such circuits.

The vertical output amplifier operates in Class A. A linear sawtooth voltage in the input creates a linear sawtooth current in the deflection coils (Fig. 8-5). Either a PNP or NPN transistor can be used in the vertical output. It must be able to handle a peak-to-peak current of about 500 milliamps. This is about the normal current.

An NPN transistor requires a positive-going sawtooth pulse. A spike of collector voltage is developed during the quick retrace of the sawtooth. This spike, if it were negative, would, during that match, forward bias the collector of the

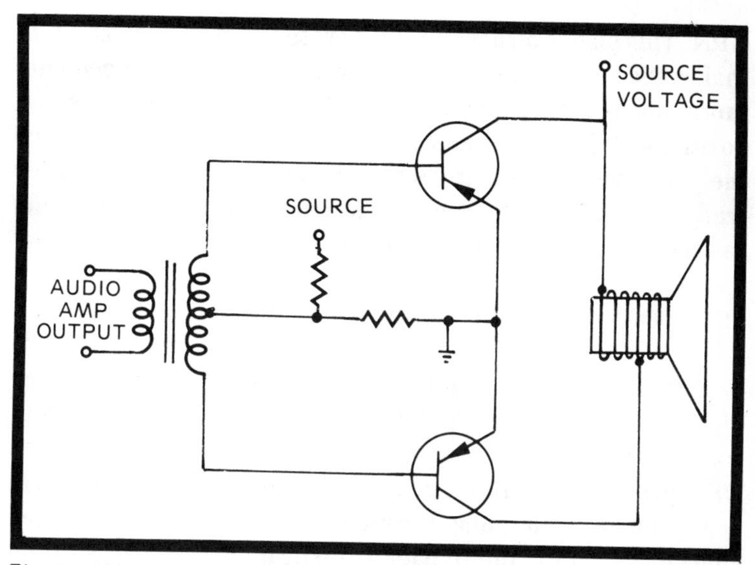

Fig. 8-4. When the output transistor impedances match the voice coil, an output transformer is not needed.

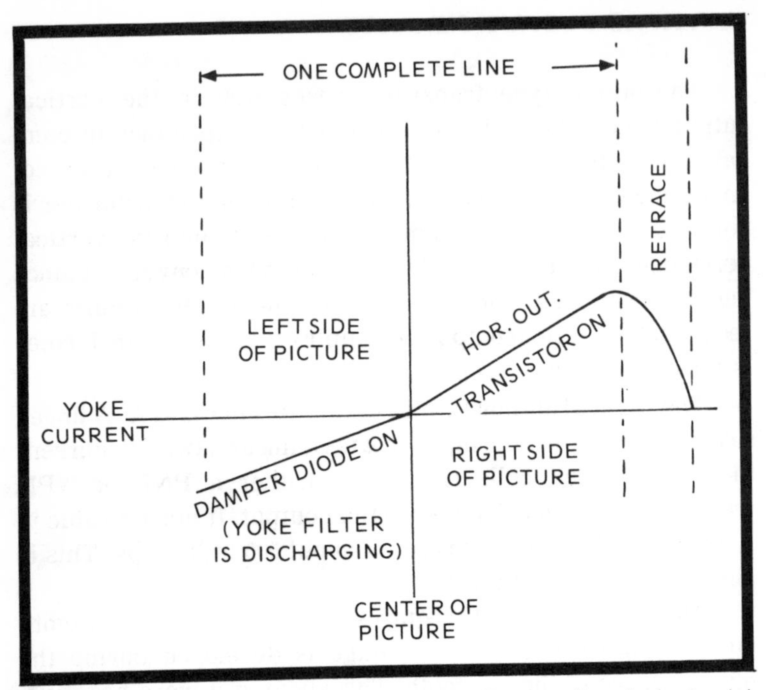

Fig. 8-5. This yoke current waveform illustrates the operation of a horizontal output stage.

NPN. This places a short circuit across the deflection coil that makes the retrace distort and slow down. The sweep becomes more linear and a distorted picture results. However, a positive-going pulse produces a positive spike in the NPN and the collector retains its desired reverse bias. If a PNP transistor is used the opposite is true and a negative-going sawtooth must be used in the input.

HORIZONTAL OUTPUT CIRCUITS

The horizontal output circuit also is the first stage of the high-voltage supply. (See Chapter 15.) The horizontal output transistor is a heavy duty type and is found typically in an NPN grounded emitter circuit. The power is developed by a heavy current coursing through the collector load. In the collector circuit is the flyback transformer, the yoke and the damper. The horizontal output circuit functions as a switch rather than an amplifying device. A square-wave signal is applied through a small transformer to the base of the output transistor. The switch circuit goes on and turns off.

The cathode ray is in the center of the CRT when the switch is off, or in the center of its sweep. The square wave enters and turns on the switch by exceeding the base's cutoff bias. A current flows into the yoke. As the current begins to flow and reaches a good strong flow, the induced magnetic field pulls the cathode ray from screen center to the right-hand side. During this half of the sweep the capacitor from collector to ground charges. As the sweep reaches the right side, the square wave turns off the switch. The magnetic field collapses, producing about 600 volts positive on the collector. During the collapse the cathode ray is snapped back to the left side of the screen as the field reverses itself.

The capacitor in series with the yoke, which charges during the first half of the sweep, starts discharging through the yoke. The electrons pass to ground through the damper diode. The discharge current induces a field that begins the sweep and pulls the cathode ray to the center of the screen. At that time the switch turns on again and continues the sweep to the right side. The sequence continues over and over.

The transistor is on only during the center-to-right-side scan. The transistor is off during the retrace and the left-hand scan. The left-hand scan pulls the beam from a negative left to screen center zero. The right-hand scan pulls the beam from zero to a positive right-hand position. The current flow during the left-side scan is derived from the discharging yoke capacitor to the collector. The current flow during the right-side scan is supplied by the collector, through the yoke, which charges the capacitor.

When the left-hand part of the TV picture is poor, it could be due to a defective damper diode on the associated components. Right-hand defects indicate trouble in the switch circuit, since that's the only time it's turned on.

TRANSMITTER RF POWER AMPLIFIERS

An RF power amplifier is found in transmitters and is the last stage coupled to the transmitting antenna. Like other

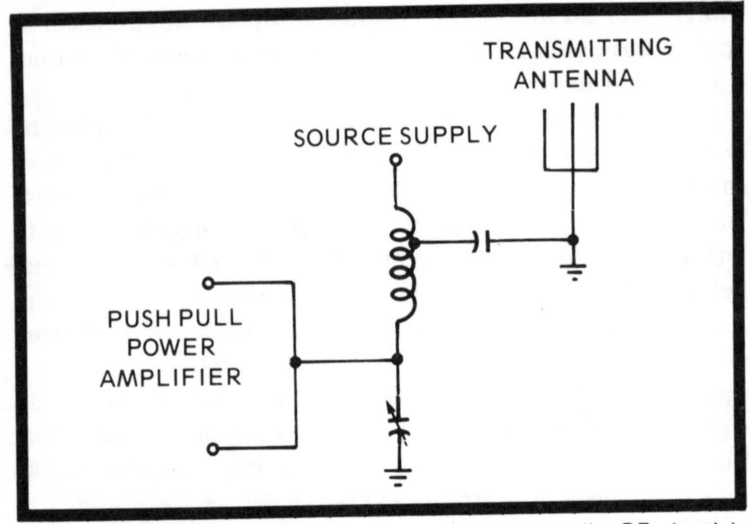

Fig. 8-6. The power amplifier in a transmitter changes the RF signal to electromagnetic waves radiating from the antenna.

power amplifiers, the output device is designed for heavy duty applications and can provide a lot of current output. The work output, of course, is the production of electromagnetic energy. The load is coupled inductively. In tube circuits the output transformer is in the output circuit and the transformer matches the output to the load, which in a transmitter is the antenna. Various configurations are used according to the actual coupling, frequency, etc. A major consideration is that the RF power output must be tuned to the frequency it is transmitting. This is usually accomplished by making the primary of the output transformer a tank circuit that resonates at the required frequency. The secondary is then a stepdown winding that matches the high tube output impedance with the low impedance of the transmitting antenna. The antenna, if it is a half-wave dipole, has an impedance of about 72 ohms.

In a transistor RF power output, the transistor acts like the tube, except for its impedances. Its output impedance is about 10 ohms. With a 72-ohm antenna a large mismatch occurs. A transformer that steps up the output from the 10 ohms of the transistor to the 72-ohm antenna is difficult to use, because the tank circuit would resonate strongly generating a lot of unwanted harmonics. Therefore, the transistor has to be matched with a tuned capacitive-inductive network, rather than a transformer, and the network has to achieve the match without producing any harmonics. The easiest way to match the output to the antenna load is with an L network, which is an inductance in series and three tunable capacitors, two in series and one in parallel to ground (Fig. 8-6). By using variable capacitors, the output network can be tuned to the transmitted frequency and also to match the load.

The RF power amplifier is typically operated in Class C, where the bias is set so the device conducts only during a tiny portion of the signal swing. This causes an impossible distortion of the signal, but it is fine when the signal is just excitation from an oscillator. The waveshape is unimportant. Only energy at the assigned frequency must be passed. In Class C operation, a suitable output is developed nicely with just a small portion of the total signal. The rest of the input signal has no use. The Class of amplifier operation can be varied from C to B by addding an emitter resistor and varying its value. This changes the EB bias to order.

An RF power amplifier is a straightforward amplifier that changes a peak-to-peak voltage into a peak-to-peak current that flows in the load. In fact, all power amplifiers do this type of job. They change the input voltages to output currents in inductances. The current that flows in the inductances induces a varying magnetic field. In the audio output the magnetic field physically moves the speaker cone. In the vertical output the magnetic field moves the cathode ray up and down. In the horizontal output the magnetic field moves the cathode ray from side to side. In an RF power amplifier the magnetic field produces radiation from an antenna which travels through the air at the speed of light.

TESTING POWER AMPLIFIER CIRCUITS

Power amplifier stage tests involve routine DC voltage and resistance measurements. Troubles in these stages are easily localized. Audio output troubles are located easily by routine signal tracing and injection techniques as discussed in Chapter 7.

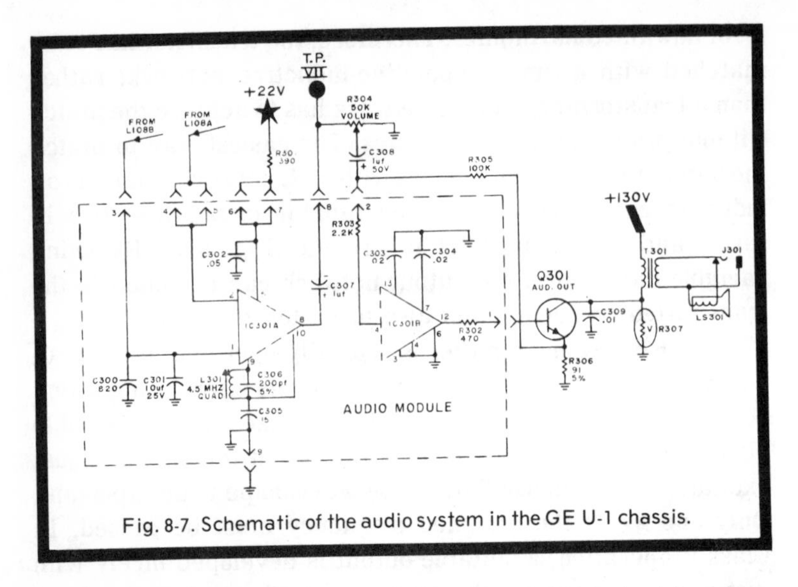

Fig. 8-7. Schematic of the audio system in the GE U-1 chassis.

Vertical output, horizontal output and RF output troubles can be localized by one scope test. The scope is set for the operating frequency, 60 Hz for vertical, 15,750 for horizontal, or the transmitter frequency, and connected to the base or gate of the output device. If there is no input waveform, the oscillator or the stages preceding the power stage are

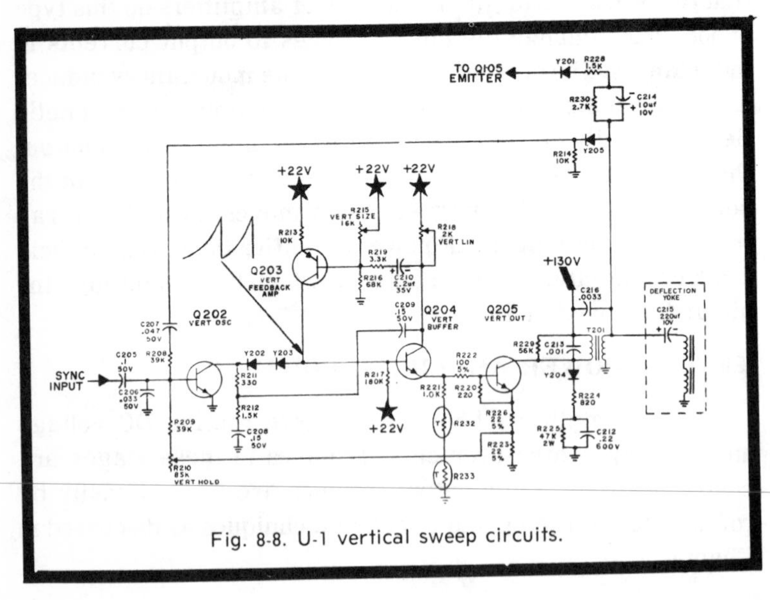

Fig. 8-8. U-1 vertical sweep circuits.

defective. If the correct waveform is present the power stage has the defect.

Output transistors are heavy-duty types and as such usually are mounted in a socket or simply holes in the chassis and held in place by two small hex head self-tapping screws. Such transistors are easily removed from the circuit. When there is a power amplifier defect, the transistors should be tested after a visual inspection for charred resistors or capacitors. The usual go no-go tests with an ohmmeter can be deceiving. Lots of transistors have reverse resistances that are lower than what would seem to be normal. Yet they are still good. The best way to test a transistor out of circuit is with a good transistor tester. Leakage is a serious defect, since transistors work with such low impedances and at various frequencies.

When replacing output transistors, be very careful that the insulators are in place and that both sides of the insulator are covered well with a silicone grease for better heat conduction. If heat sinking or metallic fins are used, be sure to replace them exactly, or you'll risk premature burnout of the transistor. Output transistors are quite expensive and mistakes are costly. Figs. 8-7, 8-8 and 8-9 are schematics of the audio output, vertical sweep and horizontal sweep circuits in GE's U-1 chassis. The following charts list troubleshooting procedures for the U-1 chassis.

(See following pages; schematic on page 144.)

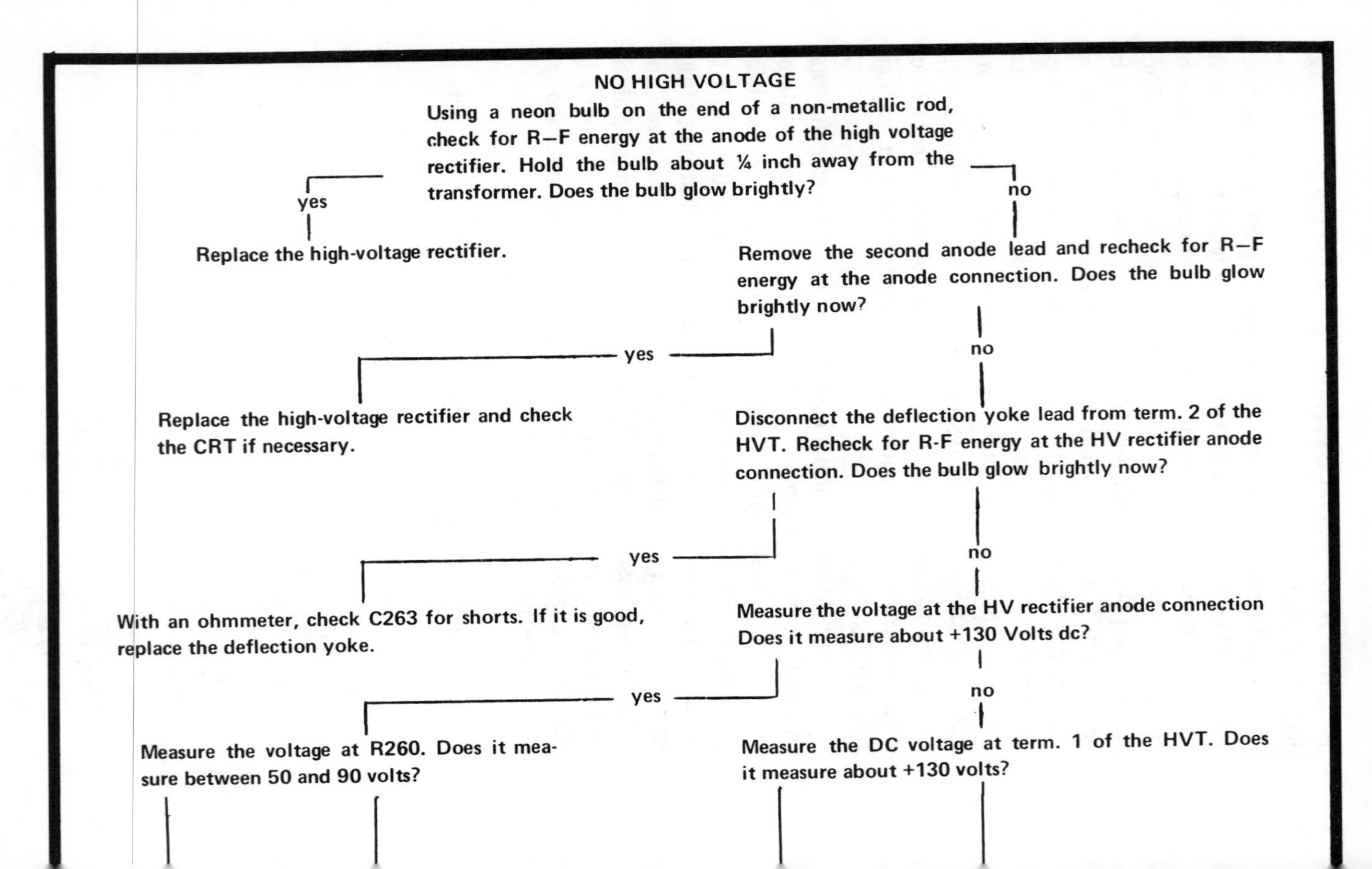
NO HIGH VOLTAGE
Using a neon bulb on the end of a non-metallic rod, check for R–F energy at the anode of the high voltage rectifier. Hold the bulb about ¼ inch away from the transformer. Does the bulb glow brightly?
yes
no
Replace the high-voltage rectifier.
Remove the second anode lead and recheck for R–F energy at the anode connection. Does the bulb glow brightly now?
yes
no
Replace the high-voltage rectifier and check the CRT if necessary.
Disconnect the deflection yoke lead from term. 2 of the HVT. Recheck for R-F energy at the HV rectifier anode connection. Does the bulb glow brightly now?
yes
no
With an ohmmeter, check C263 for shorts. If it is good, replace the deflection yoke.
Measure the voltage at the HV rectifier anode connection Does it measure about +130 Volts dc?
yes
no
Measure the voltage at R260. Does it measure between 50 and 90 volts?
Measure the DC voltage at term. 1 of the HVT. Does it measure about +130 volts?

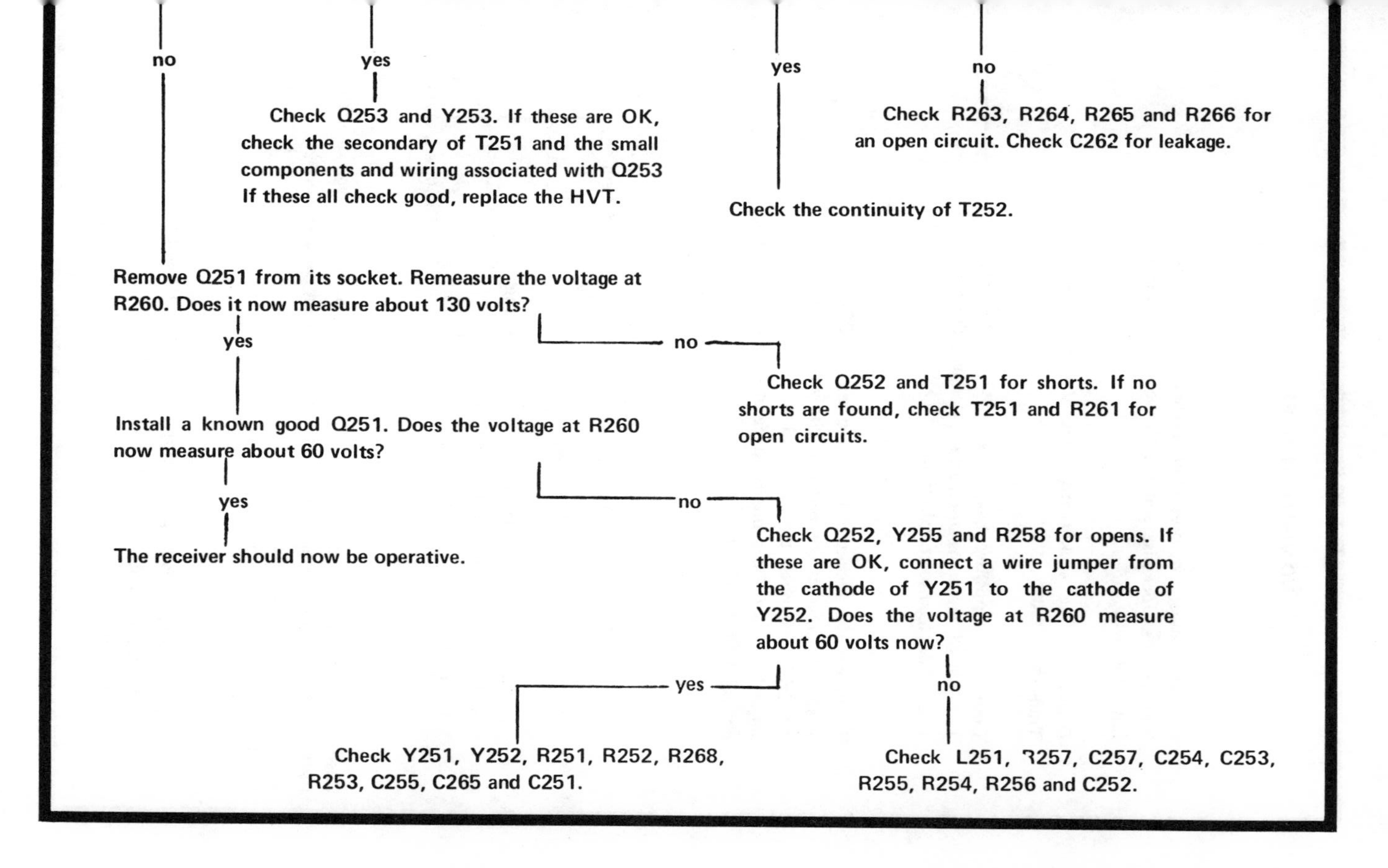
no
yes
yes
no
Check Q253 and Y253. If these are OK, check the secondary of T251 and the small components and wiring associated with Q253 If these all check good, replace the HVT.
Check R263, R264, R265 and R266 for an open circuit. Check C262 for leakage.
Check the continuity of T252.
Remove Q251 from its socket. Remeasure the voltage at R260. Does it now measure about 130 volts?
yes
no
Check Q252 and T251 for shorts. If no shorts are found, check T251 and R261 for open circuits.
Install a known good Q251. Does the voltage at R260 now measure about 60 volts?
yes
no
The receiver should now be operative.
Check Q252, Y255 and R258 for opens. If these are OK, connect a wire jumper from the cathode of Y251 to the cathode of Y252. Does the voltage at R260 measure about 60 volts now?
yes
no
Check Y251, Y252, R251, R252, R268, R253, C255, C265 and C251.
Check L251, R257, C257, C254, C253, R255, R254, R256 and C252.

NO VERTICAL SWEEP

1. With the power off, preset the Vertical Lin. control fully counterclockwise and the Vertical Size control fully clockwise.

2. Disconnect the socket from the Vertical Output Transistor, Q205.

3. Connect the positive lead of a VOM to the base terminal of the transistor socket. Connect the negative VOM lead to chassis ground.

yes

Reconnect the socket of Q205 and remove Q204 from its socket. Connect one end of a wire jumper to pin 8 of the CRT and touch the other end to terminal 3 of the deflection yoke. Does this produce at least 6 inches of vertical deflection?

yes — Touch the wire jumper to the base of Q205. Does this produce at least 6 inches of vertical deflection?

- yes — Check Y205, R214, C207 and R208. If these are all good, check Y204, R224, R225, C212 and REPLACE Q205.
- no — Measure the collector voltage of Q205. Is there any measurable voltage?
  - yes — Check Y204, R224, R225, R223, R226, C212, and Q205. IF ALL CHECK OK, REPLACE Q205.
  - no — Check T201 and the 130 volt source.

no — Touch the wire jumper to terminal 1 of the yoke. Does this produce at least 6 inches of vertical deflection?

- yes — Replace C215.
- no — Replace the yoke.

4. **Apply power to the receiver and adjust for minimum brightness and contrast to prevent burning the CRT screen.**

5. **Rotate the Vertical Hold control knob from one end of its range to the other. Does the meter reading vary from about 6.5 volts to 8.5 volts as the knob is turned? (If the voltage is less than 2 volts, check Q204, Q202, R217 and R218.)**

no

Rotate the Vert. Size control knob. Does the meter reading vary as the knob is turned?

no

yes

Check Q202, Y202, Y203, R209, R210 and C206.

Rotate the Vert. Lin control knob. Does the meter reading vary as the knob is turned?

yes

no

Check Q203, Q204, R215, R216, and R213.

Check Q204, R217, R218, R222, R220, R223 and R226.

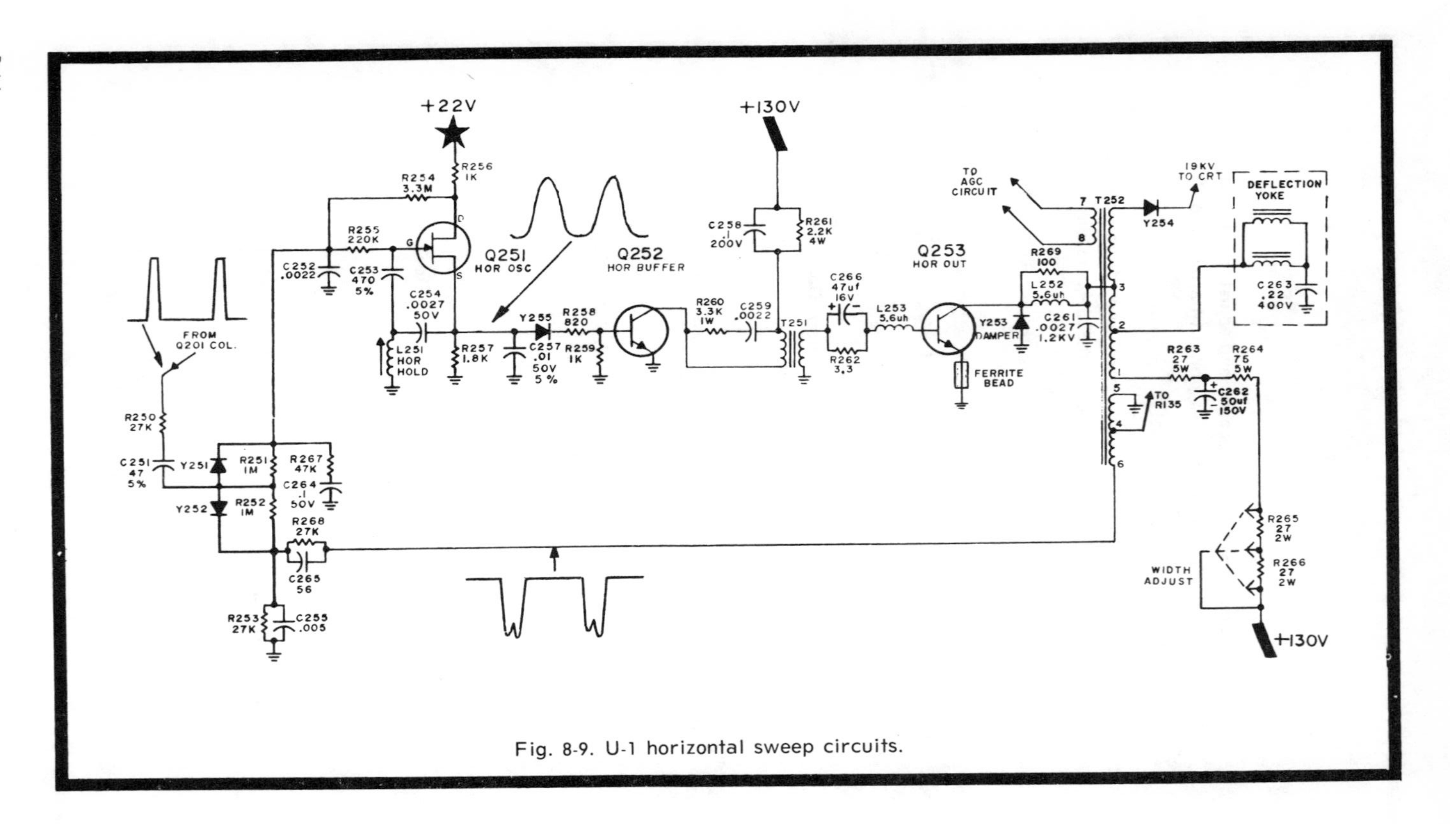

Fig. 8-9. U-1 horizontal sweep circuits.

CHAPTER 9

Oscillators

An oscillator is an electronic circuit that turns off and on at a continuous rate. The number of times it turns on and off per second is called its frequency. Frequency used to be measured in cycles per second, but that term has been changed to Hertz.

The original oscillator in electronics was used as the signal generator in a transmitter. Such oscillators are designed to run at a desired frequency and the output, after amplification, feeds current to the antenna. Other oscillators in common use are the local oscillators in receivers used to heterodyne with the incoming signal, vertical and horizontal oscillators in TVs, 3.58-MHz color TV oscillators, oscillators to drive remote control objects, and parasitic unwanted oscillation breakouts in electronic circuits.

An oscillator's frequency is controlled by a quartz crystal, by a relationship between a capacitor and inductance, by a relationship between a capacitor and resistor, or by invisible distributed capacitance, inductance and resistance between device elements, wiring and chassis ground.

CRYSTAL OSCILLATORS

It has been found that a crystal will oscillate at a relatively fixed frequency if it is placed in an electronic circuit. The frequency at which it oscillates is a direct function of the crystal thickness.

The crystal is placed in series with the feedback part of the circuit. It can be connected collector to base, emitter to base, or base to ground. A crystal has a current rating and should be run as cool as possible. The current that passes through it is on the order of a low of 10 and a high of 300 ma. If the current becomes too high, the crystal heats and the

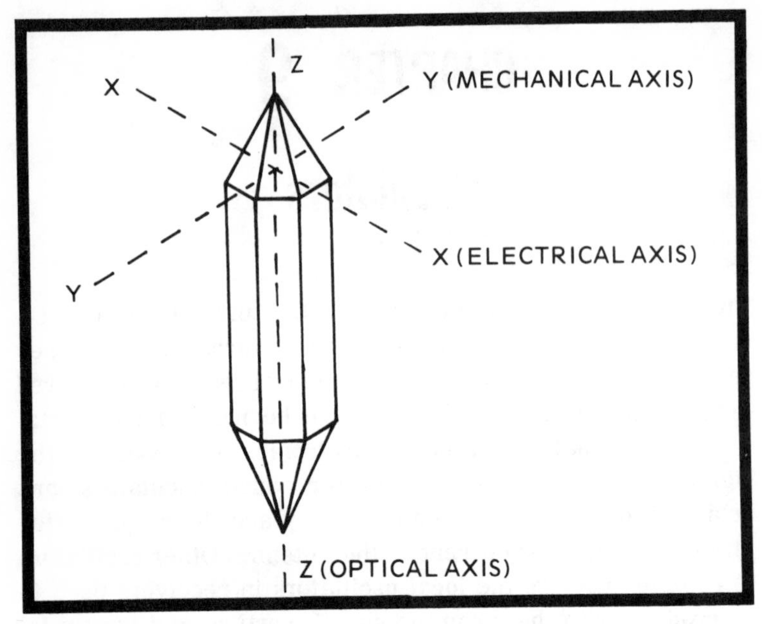

Fig. 9-1. A quartz crystal has three axis. For oscillation purposes, it can be cut anywhere in the X and Y ranges.

frequency can drift from its prescribed operating point. Too much current will crack the crystal.

In its natural state, a quartz crystal is shaped like a hexagonal prism, with each end coming to a hexagon point. The end-to-end or optical axis is named the Z axis (Fig. 9-1). From a flat side directly through the crystal to the center of another flat side is the mechanical axis, named the Y axis. From a point side through the crystal to the opposite point is the electrical axis called the X axis.

If an AC voltage is impressed across the X axis the crystal will expand and compress. It vibrates mechanically to the frequency of the voltage. Depending on the thickness of the crystal, it will have a resonant frequency. If the resonant frequency is applied, the crystal will maintain that oscillation with practically no deviation. Better than an accuracy of plus or minus 1 MHz. Oscillator crystals are cut in jewelers style along either the X or Y axis. The crystal is not voltage-sensitive along the Z axis, which, incidentally, is called the piezoelectric effect.

As mentioned before, a change in temperature can cause the frequency to vary slightly. Crystals have a temperature coefficient measured in MHz per 1 degree centigrade changes. X axis crystals have a negative temperature coefficient while Y axis crystals have a positive temperature coefficient. Heating of an X axis crystal lowers the frequency, while in a Y axis it raises the frequency. If a crystal is cut between the X and Y axis, called an XY crystal, it will tend to have no frequency change if it heats up. Different crystals are found, according to particular circuit applications.

The frequency of a crystal goes up as its thickness goes down. Conversely, the frequency goes down as the crystal is cut thicker. X-cut crystals are the thickest for a given frequency.

In reality, a crystal is nothing more in a circuit than a high Q tuned circuit (Fig. 9-2). It is mounted between two metal plates that hold it snug but make allowances for the mechanical oscillations. The two plates act as a capacitor with

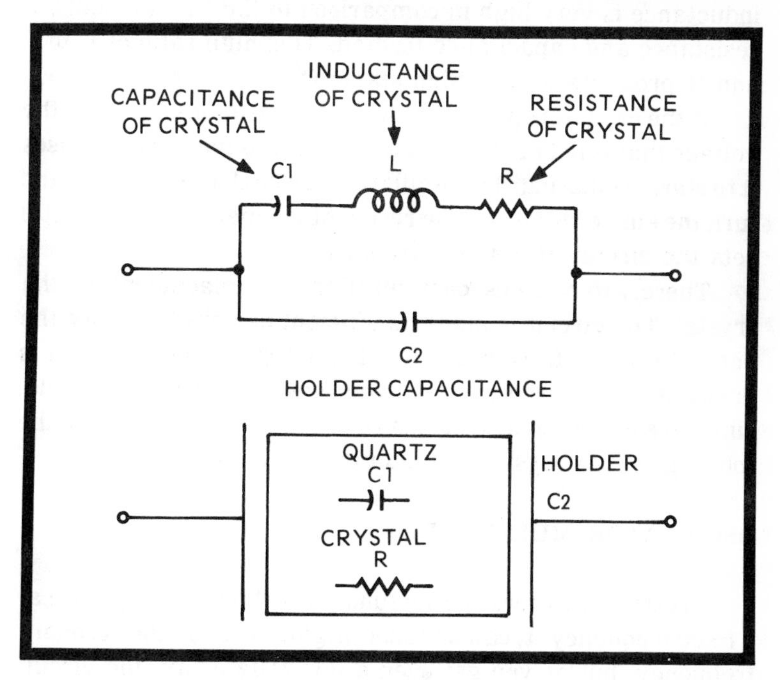

Fig. 9-2. Inherent crystal qualities include various amounts of capacitance, inductance and resistance.

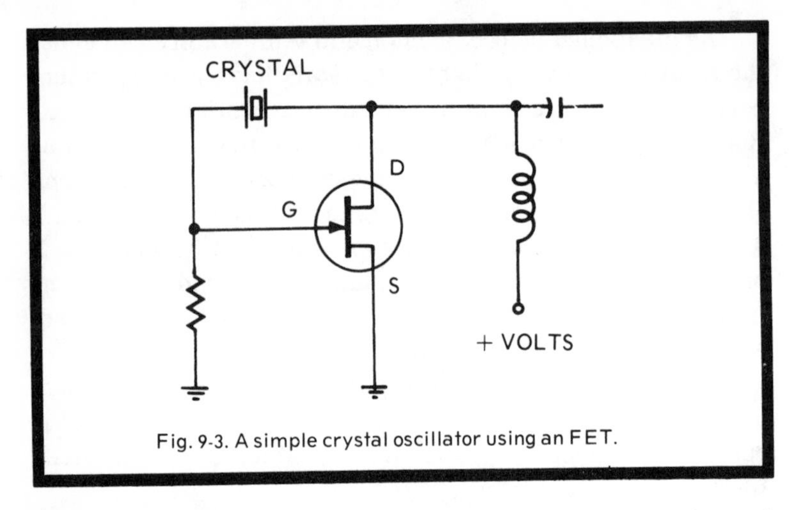

Fig. 9-3. A simple crystal oscillator using an FET.

the crystal as the dielectric. In shunt with this formed capacitor is the internal workings of the crystal itself. Inside there is a resistor, an inductance and another capacitance in series. Together they form a tank circuit. It turns out that the inductance is very high in comparison to the tiny amounts of resistance and capacitance formed. This high ratio of L to C and R produces a very high stable Q.

If the crystal is placed in the collector-to-base circuit, the voltage that is fed back from the collector to the crystal causes it to start oscillating. The oscillations go back into the base and turn the entire stage on. The resonant frequency of the crystal sets the circuit at a stable frequency.

There are various configurations for placement of the crystal. The circuit is made as efficient as possible, since the better the amplification the less the amount of current that is dragged through the crystal. Wherever an extremely stable, single frequency is needed, a crystal oscillator usually gets the job. Fig. 9-3 is a crystal oscillator circuit using an FET.

OSCILLATOR MULTIPLIERS

Crystal oscillators are not tunable; rather, they operate at a fixed frequency. It can be tuned slightly around the resonant frequency, but as you get even a few MHz away, the circuit efficiency falls off steeply. In addition, the frequency of a crystal is limited by its thickness. As you get up past 20 MHz, the thickness of the crystal becomes too thin to be practical. There are some specially made crystals that can operate much higher, but they are expensive and rare.

In order to produce crystal-controlled high frequencies, a crystal is cut for use at an easy low frequency and then harmonics of that basic frequency are amplified and used for the various jobs in transmitters, marker generators, etc.

The crystal itself is not rich in harmonics. It tends to produce only its resonant frequency. The crystal, then, is simply put into the base circuit where it resonates at the natural frequency. The transistor is biased heavily at Class C. This produces extreme distortion in the collector. This distortion creates every harmonic possible at the crystal output. Harmonics as high as 100 are easily obtained.

The desired harmonic appears in the output across a tank circuit tuned to the harmonic frequency. The tank circuit is the output load. The high stable oscillator output at a specific harmonic is thus produced.

LC OSCILLATORS

It was noted that a crystal oscillator is started by feeding back some signal from the output of an amplifier to its input. An LC oscillator operates under about the same principle, except that the feedback is an actual LC discrete component network, rather than a crystal with the LC characteristics.

A fundamental LC oscillator circuit using an FET has a tuned circuit in the gate to ground circuit (Fig. 9-4). The L part of the tuned circuit is part of a transformer. The other winding is in the drain circuit in series with the DC potential from the power supply.

As the circuit is energized, electrons flow from the source to the drain and then through the drain coil to the power supply. As the current flows, the drain coil is energized and induces an equal but opposite current in the gate coil. A voltage begins to build across the tank circuit as the capacitor charges. As the charge accumulates, the gate is turned on stronger, which makes the drain current increase, charging

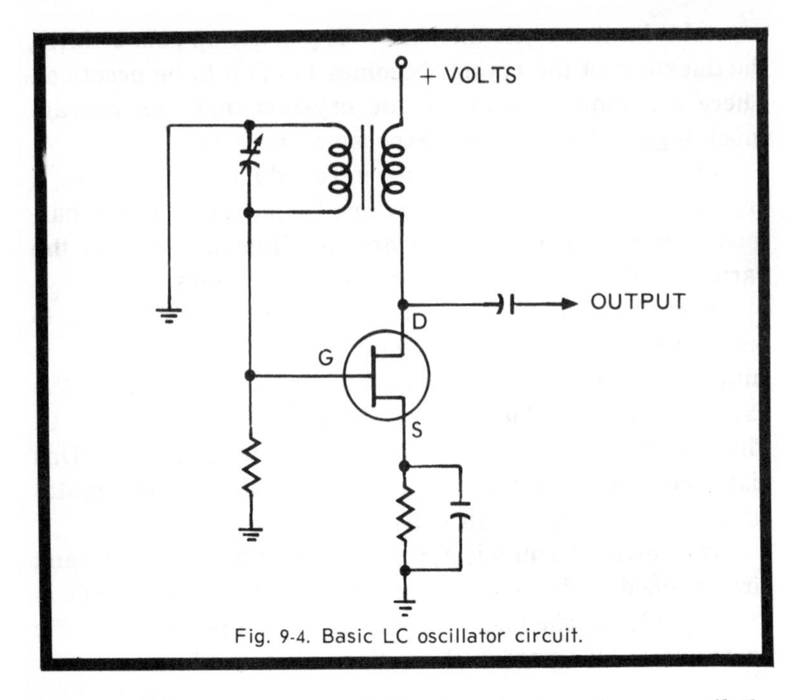

Fig. 9-4. Basic LC oscillator circuit.

the capacitor even more. This charging continues until the FET saturates. At that point, the magnetic field around the drain coil stops increasing. The capacitor, with no more current flow into it, starts to discharge. This tends to turn off the gate, which turns off the drain current.

The drain coil field collapses, and the collapse causes another magnetic field to be induced in the gate coil but opposite to the original field. This turns off the FET. Electrons flow in the other direction in the gate coil, charging the capacitor with the opposite polarity. The capacitor starts discharging from the other direction. The gate starts to turn on the FET again and the drain coil begins to develop its original type magnetic field and the cycle continues.

The LC effect of the tank circuit in the gate causes the capacitor to charge and discharge at its resonant frequency. If the capacitor is variable, the circuit can be tuned. (A crystal oscillator can't be tuned easily.) It should be noted that in an LC oscillator, the feedback produces a signal on the gate that is in phase with the drain signal. The phase change takes place in the small coupling transformer from primary to secondary.

Another way in-phase feedback is accomplished is by feeding the output signal to the source of an FET or the emitter of a transistor. In these devices, like a triode tube, the control grid, gate or base has a signal that is naturally 180 degrees out of phase with the plate, drain or collector. The cathode, source or emitter signal is in phase with the plate, drain or collector.

In a typical transistor oscillator (Fig. 9-5), there is a feedback path from the collector to the emitter, with a capacitor attached between the two connections. The signal can be developed across an unbypassed emitter resistor. The base is bypassed and the base bias determined by a voltage divider. The oscillator can be tuned with a variable capacitor and an inductance in the collector circuit. The oscillator load is across the tank and delivered to the next stage. This is a common type oscillator circuit found in black-and-white TV tuner.

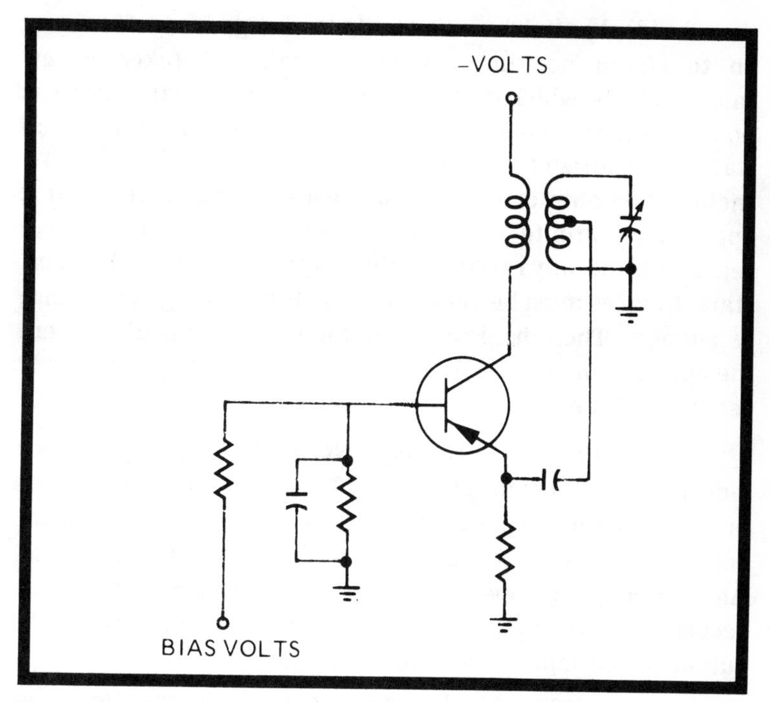

Fig. 9-5. Feedback can be applied to either the gate-source or base-emitter legs.

RC TYPE OSCILLATORS

While crystal oscillators are used where an accurate fixed frequency is required and LC oscillators used where tunable frequencies are needed, RC oscillators are used in relatively low-frequency circuits. Crystal and LC oscillators usually produce a sine-wave output, while RC oscillators produce a square or sawtooth wave. The RC oscillator is more like a switch turning on and off suddenly to produce a waveshape with sharp edges. An LC oscillator goes on and off smoothly, producing the sine wave.

As the letters RC imply, the switching action of the oscillator is produced by the time constant formed by resistors and capacitors. Transistors, FETs and silicon controlled rectifiers all make excellent switches, much better than tubes. These devices are commonly used in TV horizontal and vertical scanning oscillators.

The ideal switch has the ability to change from off to on or on to off in no time at all. Actually, it takes a few microseconds, which is satisfactory. Also, the ideal switch has no resistance in the on state and infinite resistance in the off state. A transistor approaches zero ohms in the on state (actually an ohm or two) when it goes into saturation and it approaches infinite resistance in the off state. Its leakage represents the only resistance that might interfere with the off state. Devices must be chosen with as little leakage resistance as possible. Then the device is put into a configuration where the application of DC voltage starts the circuit oscillating. It oscillates at the frequency determined by the RC components.

The most common transistor RC oscillator is the familiar 2-device multivibrator (Fig. 9-6). Typically, the two emitters are tied together. One coupling capacitor connects from the collector of the first transistor to the base of the second. Another coupling capacitor connects from the collector of the second to the base of the first. This feeds back some of the output to the input to sustain oscillation.

Lastly, there are the usual base leak resistors and collector load resistors. There are variations, but that is the basic circuit. It works like this: First, the frequency is determined by the time constants of the feedback C and the first base leak R, the coupling C and the second base leak R, and the cutoff points of the two transistors. A proper combination provides such divergent frequencies as 60 Hz or 15,750 Hz for vertical and horizontal TV scanning frequencies.

As the DC voltage is applied, one of the transistors will produce slightly more collector current than the other. Let's suppose it is the first transistor. That transistor conducts heavily. This charges the coupling capacitor and the charge begins leaking off through the second base leak resistor.

When the entire charge leaks off, the second transistor reaches a turn-on bias. The second transistor conducts and charges the feedback capacitor, which cuts off the first transistor. The feedback capacitor is feeding an in-phase signal from the second output to the first input. (The signal "turns" 180 degrees through each transistor and there are two transistors, "turning" the phase 360 degrees.) The charge leaks off through the first base leak resistor and the first transistor turns on. This turns off the second and the cycle keeps repeating itself.

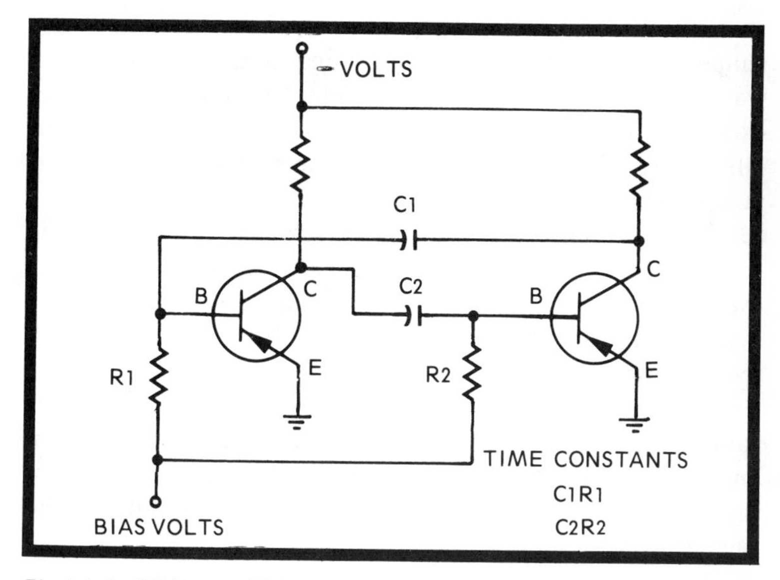

Fig. 9-6. An RC type oscillator uses no inductance. The frequency depends on the CR time constants.

TESTING OSCILLATOR CIRCUITS

Basic oscillator tests require answers to two questions. One, is it oscillating? Two, if so, what frequency is it on? Once these questions are answered, the actual DC operating characteristics and components can be checked routinely.

If the oscillator is running within the range of your oscilloscope, or even a bit higher, these questions can be quickly answered. The scope probe is touched to the output of the oscillator stage. The waveshape (or lack of it) will appear. Its amplitude represents the peak-to-peak voltage of the oscillator output. The frequency of the waveshape is automatically determined by the setting of the horizontal sweep on the scope.

Many scopes have a range of only about 200 kHz, but there are plenty of other shop scopes that take you up as high as 10 MHz. Even with the narrower band scopes, though, lots of oscillators can be tested without seeing the exact waveshape. For instance, the 3.58-MHz color oscillator output can be viewed on any scope. It will appear as a modulation envelope rather than a single waveshape.

Not only the frequency can be determined easily. In vertical and horizontal sweep circuits, the actual edges and shape of the wave can be analyzed and compared with waveforms included in the service notes. But higher frequency oscillators, such as those in transmitters, receivers and others, cannot be checked easily with an ordinary scope. Other procedures must be used.

In tube oscillators, there is the old DC test: Check the control grid for a high negative bias. A good negative bias means the oscillator is probably running. Unfortunately, this test has no place in transistor circuits. There are other quick tricks, however.

First of all, an oscillator radiates. In a transmitter there is radiation around the oscillator circuit right in the air. If it is of any wattage, a neon bulb will light up, however dimly, from the RF energy. Place a neon near the oscillator. If it lights, the oscillator is running.

In AM radios you can check the operation of a suspected oscillator with another AM radio. Tune the second radio to the top of the dial around 1500 kHz. Bring the radio under test near the second radio, turn them both on and tune the suspect radio. If the second radio starts chirping or whistling at all, the oscillator is running. If not, the suspect oscillator is not running.

Local oscillators can be checked with DC readings, but the test is subtle. You are going to test for any changes in bias. The voltmeter is hooked across the base and emitter. A slight bias will be read, either positive or negative, it doesn't matter which. Next, the tuning of the oscillator is varied from one end of the dial or selector to the other. If the bias varies at all during tuning, the oscillator is running. If the bias remains fixed, the oscillator is dead.

In a multivibrator, the oscillator can be running and on frequency, yet there can be serious trouble in the circuit. For instance, in TV scanning, the vertical sweep could be badly distorted, yet the oscillator is running and on frequency. A close analysis of the circuit is needed to locate the cause of such trouble. The square wave that is produced is humped and sawtoothed to produce a linear sweep. The waveshaping method varies considerably from one TV to another. It is up to you to analyze the circuit carefully, locate the capacitors, resistors and transistors that are processing the waveshape, then analyze the waveshape to see where it is distorted. From the distortion, a particular component or set of components can be indicated as the cause of the trouble.

The service notes of the actual receiver usually include waveshapes that should be present at particular test points. Actual frequencies can be determined by feeding the unknown frequency into the vertical scope input and a known frequency into the horizontal input. The frequencies should be within the capabilities of the scope. If the signals are sine waves, when they are both identical, a circle will appear on the scope. This is called a Lissajous figure. When the frequency is 2 to 1, a figure eight will appear. Lissajous figures can provide the technician with a reading of the exact frequency, depending on the accuracy of the known frequency. Any good textbook on oscilloscope applications will explain the use of Lissajous figures to determine frequency.

CHAPTER 10

Converter Circuits

A converter circuit in any receiver changes all incoming signals from the original frequency to a standard intermediate frequency for amplification. A converter stage combines mixer and local oscillator functions. The incoming RF signal is passed through the converter where it encounters the local oscillator output. The two signals mix in the amplifier and the final output is the sum and the difference of the original signal, plus the harmonics of the RF and oscillator frequencies.

In some receivers the converter stage is made up of separate oscillator and mixer circuits. In these receivers the RF signal is applied to the mixer and the oscillator signal is injected into the mixer. The mixer then produces the aforementioned output.

In a tube-type converter a special pentagrid tube is commonly used. It has two control grids. Into one of the control grids the RF signal is injected. On the other control grid the local oscillator is applied. The two signals mix in the electron stream of the tube and a multitude of frequencies appear in the plate circuit. The output circuit is tuned to the desired intermediate (IF) frequency and it is tapped off.

In solid-state circuits, ordinary transistors, FETs and even diodes are used in the conversion process. FETs turn out to be the best mixer performers. They closely resemble tubes in amplification characteristics; they reject extraneous signals, cut down on crosstalk and do not overload as easily as transistors. Dual-gate FETs can be used like pentagrid converter tubes, since they have the two gate inputs.

Diode mixers are used mostly in UHF tuners, since they have a low noise figure no matter what the frequency. Other devices produce a large amount of noise at high frequencies.

CONVERSION LOSS

The converter stage, or the mixer-oscillator, operates with an advantage over the ordinary amplifier, since it has different input and output frequencies. The inputs are the RF and the local oscillator injection. The output is a mixture of these two, usually the difference frequency, the IF. Any feedback that occurs between the input and output is inconsequential, since the input is not tuned to the IF. As a result, the stability of the stage is excellent and it can be run full blast without breaking into oscillation.

On the other hand, as the two frequencies are mixed, energy is lost. The gain of the stage drops drastically. Even though the stage is operating at maximum efficiency, its actual gain is little or none. The total result is that an incoming signal frequency is changed from the RF to the IF without any appreciable gain.

TRANSISTOR CONVERTER

A converter stage, as mentioned in the beginning, combines both the mixer and oscillator functions in one stage. The advantage to this is the saving of an extra stage. However, separate mixers and oscillators provide superior performance.

In a single transistor converter (Fig. 10-1), the RF and oscillator frequencies are both attached to the base. The oscillator coil and the IF transformer primary are both in the collector circuit. They do not interfere with each other, since they are both tuned to entirely different frequencies.

This system works well, except for one major consideration. The amplitude of the oscillator signal cannot become too large. If it does, too much collector current is passed and the collector voltage drops too far. When that happens the gain of the converter becomes so low that little or no output is produced. The oscillator coil must be tapped to produce enough impedance to sustain the oscillations, but not too much or else it will load down the converter. While base oscillator injection is common, the oscillator energy can also be developed across the emitter resistor.

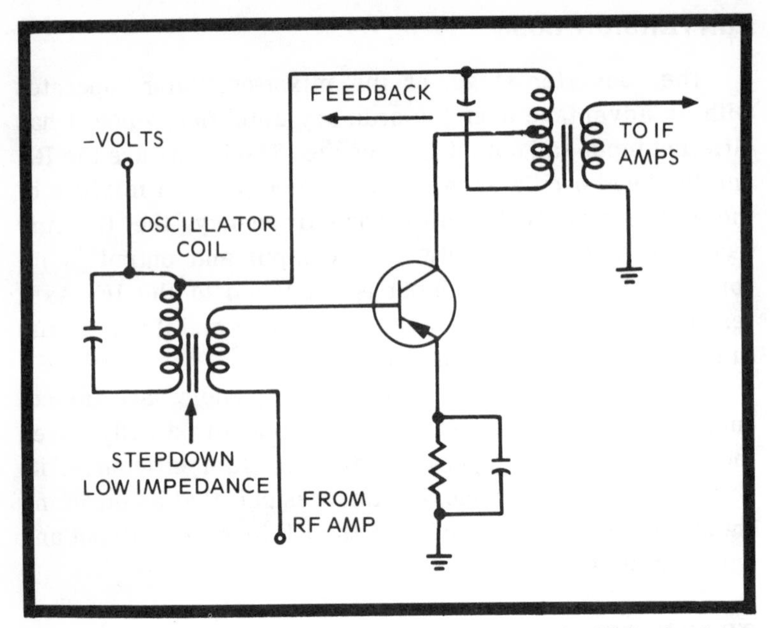

Fig. 10-1. A transistor converter has its own oscillator in the CB feedback leg.

TRANSISTOR MIXER

When a transistor is used as a mixer with a separate oscillator circuit (Fig. 10-2), considerably more gain is available from the circuit, in comparison to a one-stage converter. The RF signal is fed in through the base, since more gain is attained with the base injection than with emitter injection. It is important to amplify the RF signal carrying the modulation, rather than the oscillator signal. The only job the oscillator signal has is to help change the RF to the IF, so it needs to have only enough strength to do the job. The oscillator coil resonant point should be far from either the RF or IF frequencies. This creates a large impedance difference and the coil will not load either the RF or IF.

Bias on the base can be taken from the power supply, so the base can be as independent as possible from the oscillator signal. Even with all the design care, though, transistor mixers cannot handle high levels of signal without overloading and developing a lot of crosstalk between the frequencies. A transistor mixer has to be equipped with highly selective circuitry to avoid these problems. However, the low input impedance of the transistor creates the problems, as we have discussed.

JFET MIXER

The FET, due to its similarity in characteristics to a tube, turns out to be a better mixer (Fig. 10-3). The FET is quite like a triode tube, without the heater or size requirements. The FET circuit configuration is almost identical to the triode tube circuit.

The RF amplifier feeds the gate of the FET through a small blocking capacitor (Fig. 10-4). The FET input impedance is very high, so the oscillator injection can also be sent into the same gate connection. Or if desired, the oscillator injection can be applied at the source connection, but the bypass capacitor across the source resistor has to be removed in order to allow the oscillator signal to develop across the resistor.

In an N-channel JFET, the two signals modulate the source-to-drain current and the signals mix. The drain current, then, contains the various signals that are produced

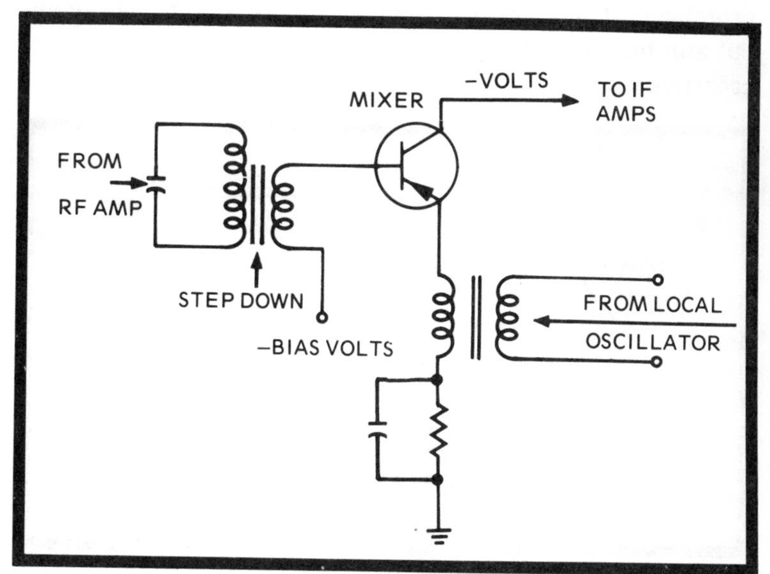

Fig. 10-2. In a typical transistor mixer stage the RF is applied to the base and the oscillator signal is injected at the emitter.

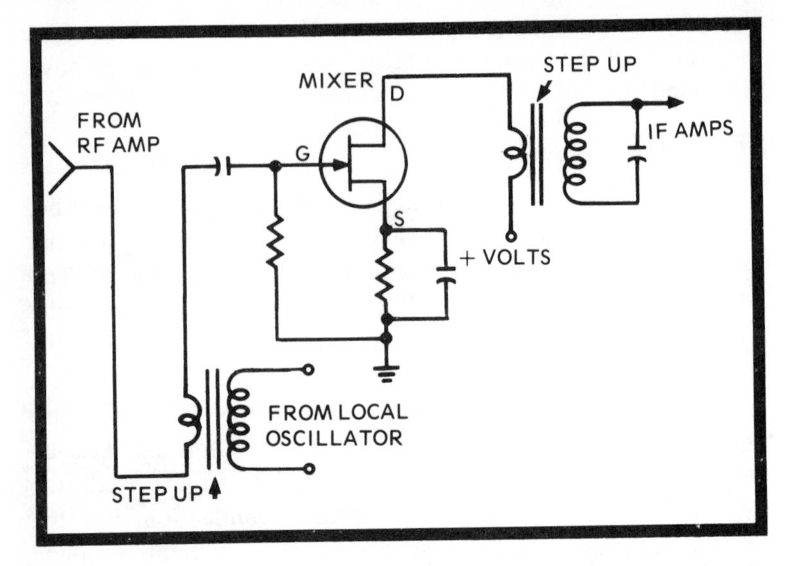

Fig. 10-3. In an FET mixer, both RF and oscillator signals are applied to the gate circuit.

by the mixing process. While the oscillator injection voltage level is kept as low as possible, the crosstalk is eliminated by raising the amount of oscillator signal injection. A compromise is struck between keeping the oscillator voltage low but still high enough to enable the conversion gain to reject crosstalk.

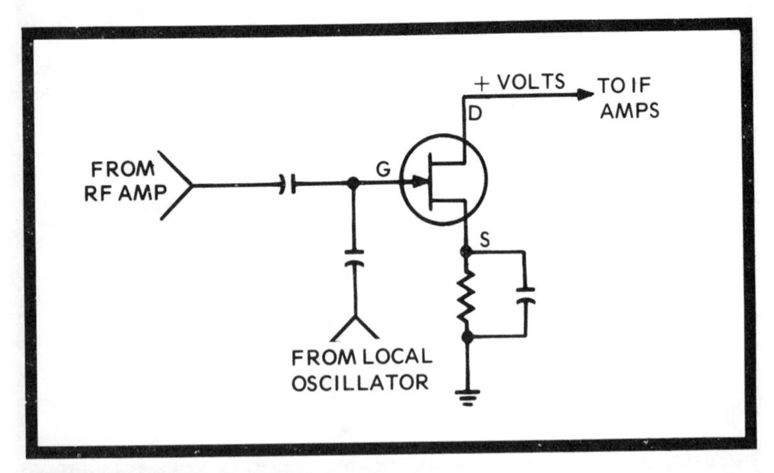

Fig. 10-4. In an FET mixer, the RF and oscillator signals can be injected through capacitors instead of inductances.

DUAL-GATE MIXER

The dual-gate FET, usually a MOSFET, is the solid-state equivalent of the pentagrid converter tube. Its characteristics are similar to the tube, and the dual gates are used just like the two control grids in the tube (Fig. 10-5). Typically, gate 1 is supplied with signal directly from the RF amplifier circuit. The input signal can come from the secondary of the coupling transformer between the RF and mixer stages. At Gate 2 the local oscillator signal is applied through a tiny coupling capacitor.

Since the source resistor is bypassed, DC flows between source and drain. The two signals entering the gates both affect the current flow and the signals are mixed and appear in the drain current. This type of mixer, like its tube counterpart, has almost complete immunity to overload and crosstalk. The electron flow is modulated as it passes each gate instead of receiving the shock of excitation all at once as it does when the modulation is applied by a single gate FET or transistor.

DIODE MIXER

Diode mixers are used in UHF applications. The ordinary amplifier has great difficulty in mixing signals as the UHF

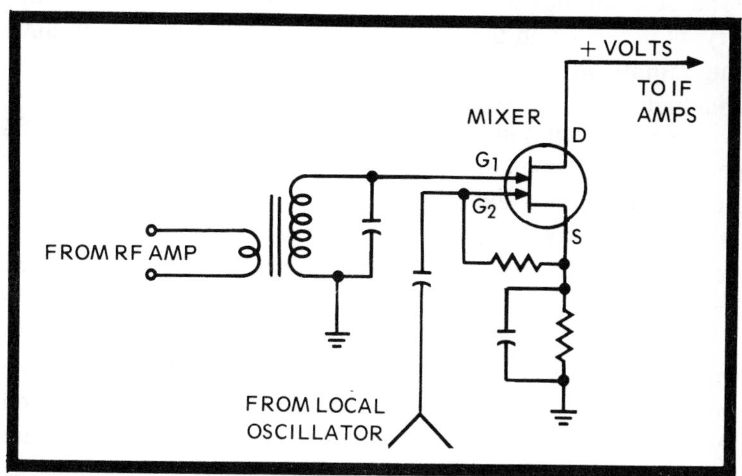

Fig. 10-5. A dual-gate FET mixer circuit is almost identical to that used with a pentagrid converter tube.

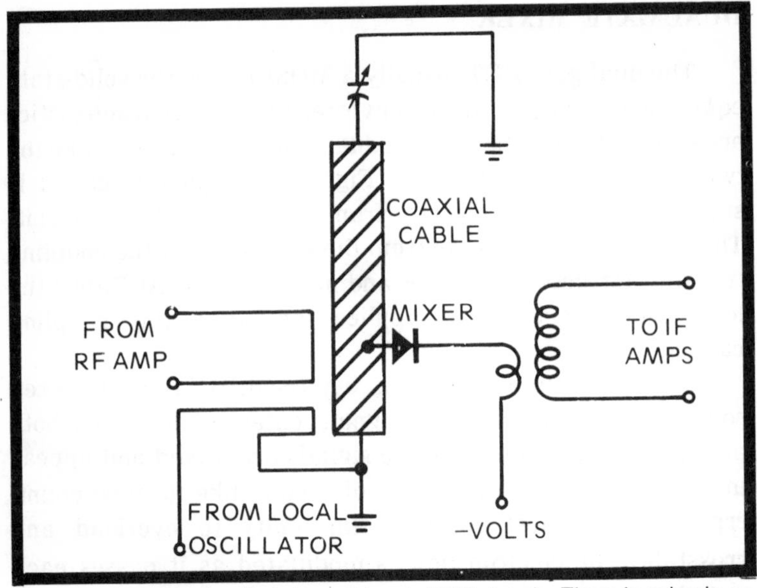

Fig. 10-6. A hot carrier diode is used to mix UHF signals. There is not only no gain, but a considerable signal loss in such mixers.

spectrum is reached. In a diode, the frequencies that are sent in also appear in the output. The simplext diode mixer, found in most UHF TV tuners, is a single hot carrier diode. A hot carrier diode is one that is strongly forward biased. In a typical UHF tuner, the forward bias can be +20 volts. This type of diode has a low noise figure in comparison to amplifying devices and the frequency response doesn't vary until the middle of the microwave region, way above UHF, is reached.

The RF signal and the oscillator injection signal are both coupled into the diode by a coaxial cable that matches the impedance almost exactly (Fig. 10-6). Of course, there is no amplification in the diode to offset the conversion loss. As a result, the diode output is always a function of the conversion loss.

A good RF amplifier should be used ahead of a diode mixer for best reception. A good, low-noise IF circuit after the mixer helps a lot, too. The smoothness of the wave produced by the local oscillator is important and the amount of local oscillator injection voltage is a large factor. Any of these factors being poor could contribute to a poor output signal.

In order to improve the performance of a diode mixer, instead of using one diode, four can be used in a balanced configuration (Fig. 10-7). They are attached in a series bridge. The oscillator injection takes place across two opposite intersections, the RF input is attached across the other two intersections, and the IF output is taken from a center tap on the RF input secondary. The injection secondary has a center tap to ground.

There is still no amplification in a bridge circuit, but the balance keeps all losses to a minimum. The IF output impedance is in the order of 72 ohms and a piece of 72-ohm coaxial cable can be used as a matching transformer to the IF input. Since the IF input is quite often a VHF tuner, it is quite convenient to use the coax between the UHF and VHF tuners.

DIRECT-CONVERSION CIRCUITS

A largely ignored heterodyne technique is called direct-conversion. It is a mode of operation that has not been common, but it will be used more and more since it lends itself well to integrated circuits. In a direct conversion circuit, the local oscillator is set to run at exactly the same frequency as the incoming RF signal. That way, when they beat together, the difference frequency is zero and the IF is zero. The modulation that is contained in the RF signal varies around the zero IF

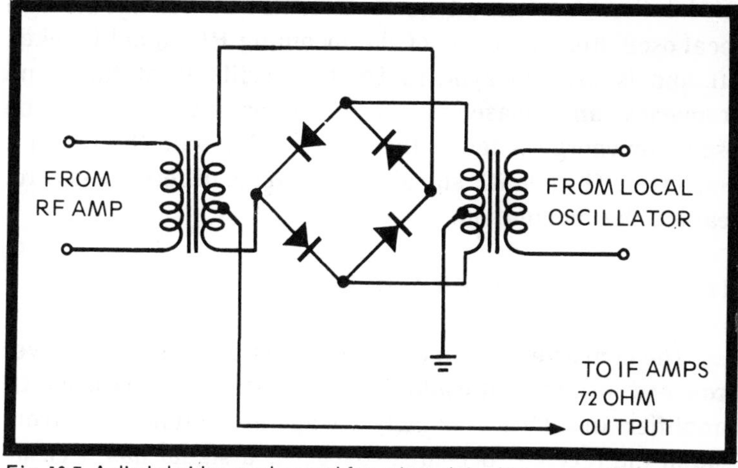

Fig. 10-7. A diode bridge can be used for mixer duty. It is more efficient than a single diode, but there is still no gain.

point. It is not only converted but detected as well. This output can be fed directly to an audio amplifier. There is no need for an actual IF amplifier or a detector. Furthermore, it is an electronic fact that the lower the IF, the better the selectivity of the stage. At zero IF the selectivity is as good as it's going to be.

The direct conversion technique has been ignored largely because there are some serious drawbacks. These drawbacks were difficult to overcome until recently. The drawbacks in order are the numerous heterodyne "whistles and birdies," much more than in any ordinary superhet, nonlinear reception over the dial, and a bad case of distortion at the slightest misalignment. The receiver works well on only high signal levels. A weak signal encounters all the above problems. However, all the new squelch circuits should make the whistling disappear. New linear circuits ahead of the converter stage can keep the signal level high enough to cure the other problems of distortion and critical signal levels.

The two direct conversion methods are called **homodyne** and **synchrodyne**, instead of the heterodyne. The homodyne is a receiver type which has no local oscillator. A portion of the incoming RF signal is tapped off, processed and then added to the converter along with the original RF signal. The two RF signals beat together and cancel each other to produce the modulation. A synchrodyne receiver (Fig. 10-8) has its own local oscillator. A portion of the incoming RF signal is taken off and is used to synchronize the oscillator at the same frequency and phase as the incoming signal. Then the oscillator energy is fed to the mixer, along with the RF in a converter. The two RF signals beat and the difference is zero, leaving the modulation.

TESTING CONVERTER CIRCUITS

The converter or mixer-oscillator is part of the receiver front end. In a radio it could be the first stage if there is no RF amplifier. In a TV it is usually a stage in the tuner. The front end of the receiver becomes suspect if a signal has been injected into the input of the IFs and is passed successfully. The next step in localizing the defective stage, using signal injection, is an attempt to pass the signal through the mixer. Signal tracing has to be stopped here since the signal level in this area is in the microvolt range and not viewable with ordinary shop scopes and probes.

Signal injection into the mixer is usually easy, since factory alignment techniques use the mixer input as an injection site. Quite often, there is a test lead coming out of the mixer input for alignment signal injection. It's perfect for the mixer localization test. Once the mixer input is found, a signal generator is set to one of the frequencies the receiver is supposed to receive and the signal is attached to this test point. Then the receiver is tuned to the generator frequency and turned on.

The oscillator will beat with the generator frequency in the mixer, if the mixer is operating. In a radio the modulated 400-Hz tone of the generator note can be heard, and in a TV the 400-Hz note will produce a number of black-and-white hum bars on the CRT face. If the mixer is not operating, no such sound or picture display will take place. When the mixer appears to be the defective stage, routine DC voltage and resistance tests are begun.

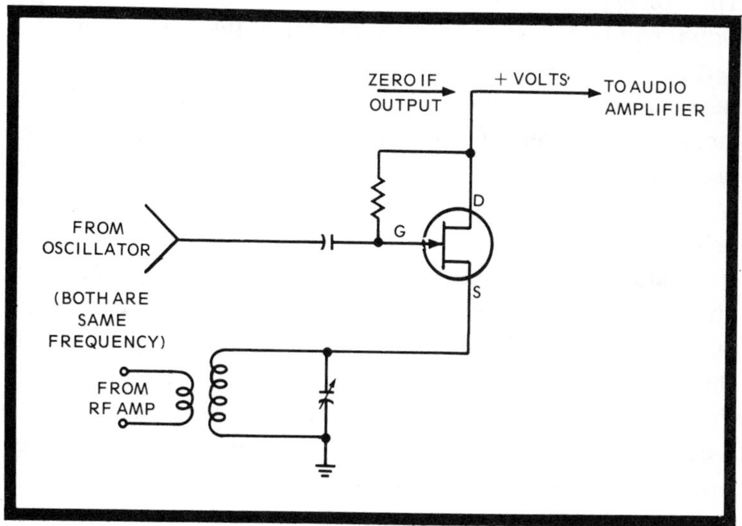

Fig. 10-8. A synchrodyne converter has an IF of zero. Therefore, it also detects automatically.

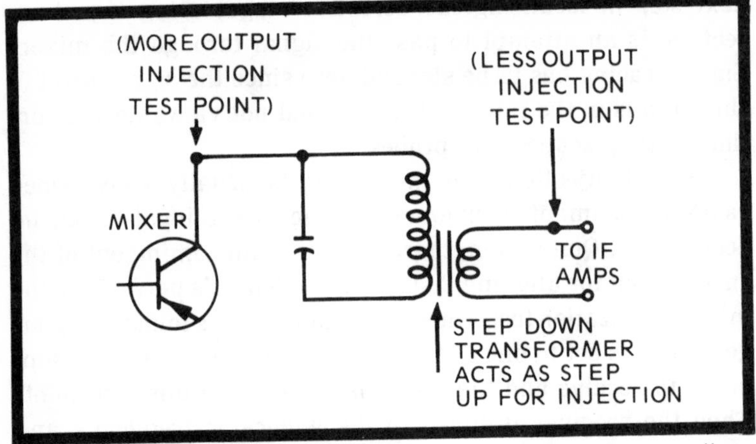

Fig. 10-9. When using signal injection tests, be careful of unfamiliar results as you pass over transformers.

TRANSISTOR COUPLING TRANSFORMERS

Signal injection in bipolar transistor circuits can be confusing to technicians who are used to tube receivers. Due to the low input impedance of transistors, the coupling transformers between stages, such as the mixer and first IF, have a stepdown ratio from primary to secondary (Fig. 10-9). Therefore, as you touch test points, different levels of output signal can be heard or seen. If you touch the first IF input, a certain level of signal is displayed. Then when you cross over the coupling transformer to the mixer output, the signal level increases considerably. Yet there is no amplification device present. Or is there?

Actually, there is voltage amplification between the primary and secondary of the transformer. If this fact of transistor signal injection is not realized, a false, confusing clue is given. The servicer scratches his head trying to figure out how in the world the output of the mixer is stronger than the input of the first IF stage. According to past experience they should be the same. The signal level will drop when injection is made from primary to secondary, as well as when injection is made from the base to the collector of the same transistor amplifier.

CHAPTER 11
Detectors

The radio frequency signal coming from a transmitter is modulated with some sort of intelligence. It could be AM audio, FM audio, video, color sidebands, etc. The received RF is amplified in the RF amplifier, changed to an IF in the mixer-oscillator, amplified more in the IF amplifiers, and then fed to some sort of detector circuit.

The detector circuit has the job of first rectifying the IF signal, getting rid of the IF and then developing what is left, which is the modulation. There are three principal considerations in dealing with detectors: the frequency being worked, the type of modulation used on the carrier wave and how true the modulation must remain after detection.

Although signals have been detected in triode and pentode tubes, detection is most commonly performed by diodes. In years past that meant diode tubes, but today the germanium crystal diode is the common device. Some receivers use transistors and even an FET as the detection device, but mostly it's the germanium diode.

A diode can be designed to exhibit excellent linear characteristics, whether a strong local signal or a weak distant one is being processed. This linear characteristic enables the diode to react in direct proportion to the strength of the signal. That way, a DC sample can be taken off and used for AGC, AFC and other type signal control voltages (see Chapter 12).

The modulated RF changes in two ways as it passes through the detector. It can enter the detector as an AM RF signal. As it passes from cathode to anode, current flows and voltage is developed across the load resistor in the cathode. Then as the RF cycle passes through zero base line and reverses polarity, the RF tries to pass from anode to cathode.

The reverse bias, as long as it is within the capabilities of the diode, permits no current to flow from anode to cathode. Nothing develops across the load resistor and the RF is rectified (Fig. 11-1).

The second step occurs across the load resistor, which has a capacitor across it (Fig. 11-2). The capacitor value, when combined with the resistor forms a time constant that filters the IF, but has a high impedance for the modulation intelligence. In other words, the modulation is developed across the high impedance, while the IF is filtered out, since the same two components are a low impedance to the modulation frequency. Therefore, little or no IF signal can be developed across the resistor and capacitor.

The capacitor does this: At the peak of each positive IF cycle, the capacitor becomes charged from the rush of electrons during the forward bias. Then as the signal reverses itself and the diode turns off, the capacitor starts leaking off the charge through the load resistor. The rush of electrons, first from the diode and second from the charged capacitor, continues in a more or less steady stream through the resistor. The voltage developed across the resistor remains at a fairly constant level. The low-frequency modulation is not affected by the capacitor because the value of the capacitor is not large enough. The modulation moves slowly and defines the outline of the filtered IF.

If the capacitor were made larger it would maintain its charge even during the longer intervals of the lower frequency. At its correct detector design value it maintains its charge for the shorter intervals of the higher frequency, but falls off in charge as the lower frequencies appear.

TIME CONSTANT

The values of the resistor and capacitor are critical to perform the detection function. The capacitor charges and then discharges through the resistor. If the capacitor is made larger, it will charge and retain its charge longer. If it is made smaller, it will not retain its charge as long. It's a matter of the volume of electrons. A large capacitor simply charges with more electrons than a smaller capacitor.

If the resistor is made larger, it presents a larger resistance to the capacitor discharge. This retards the electron flow and causes the capacitor to discharge more slowly. Conversely, if the resistor is made smaller in ohms, the capacitor can discharge through it quicker. The time constant of any RC combination can be obtained quickly by simply multiplying the resistor in megohms times the capacitor in microfarads; the answer comes out in seconds (T equals RC). In AM receivers the load resistor is typically around 470K and the capacitor is around 100 pf. This gives a time constant of around 47 microseconds.

The AM detector is fairly simple and is used for sound or video applications. If you notice that the diode is positioned either one way or the other, that is, with the anode and cathodes reversed, it works fine in either direction. The only reason for the polarity variation is to provide the desired polarity for the AGC or other control voltages. In one way a positive AGC is developed and the other way a negative AGC is developed (see Chapter 12). In Fig. 11-3, the BC junction serves as a detector diode. The EB junction provides AGC voltage.

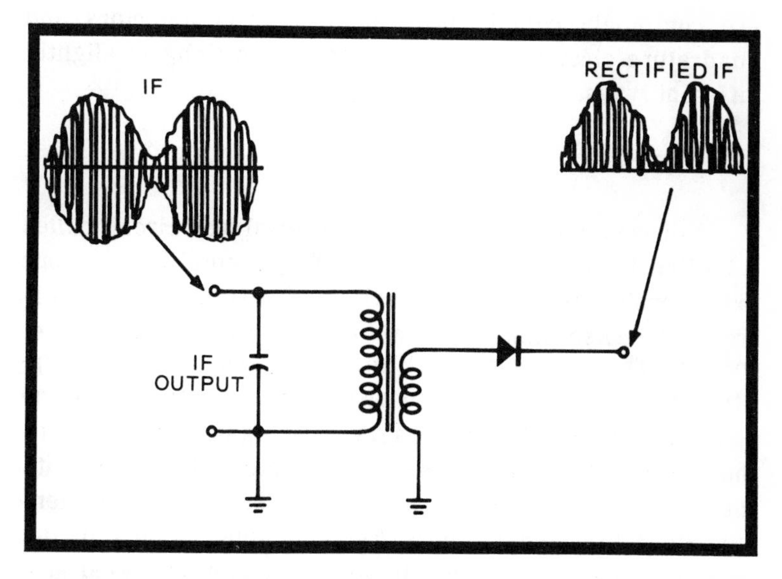

Fig. 11-1. The diode detector simply rectifies the IF modulation envelope. It does not detect.

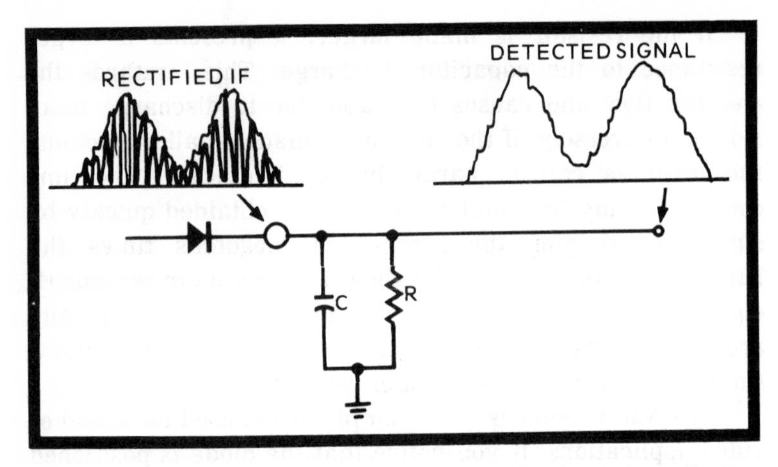

Fig. 11-2. The actual detection takes place in the RC network due to the time constant relationship.

FM DETECTORS

FM detectors perform the same job of extracting the modulation, but since the modulation is changing the frequency rather than the amplitude, an entirely different circuit is used. Two diodes are usually needed. FM detectors are commonly called discriminators, ratio detectors and quadrature circuits. They all do the same thing in slightly different ways.

Discriminator

A discriminator detector uses a special feed circuit called a **limiter**. The limiter is an ordinary IF type amplifier with one extra consideration. Even though the FM signal is relatively free of the noises that plague AM signals, some noise does get through. The noise is in an AM mode and appears in the amplitude of the FM signal. The discriminator detector has no way of eliminating the spurious AM noise signals. If they get to the discriminator circuit, they are detected right along with the FM signal. The limiter limits noise pulses by clipping them out. Since the modulation is FM and appears only in the carrier frequency changes, the amplifier can be biased at a small portion of its curve far from cutoff and near saturation.

As the signal swings from peak to peak, it alternately throws the amplifier into saturation in the positive direction and cutoff in the negative direction.

The peak positive voltage causes the transistor to saturate, producing a flat positive collector current. In other words, as the peak pushed the transistor into saturation, the peak is clipped off. The noise pulses are riding on the peak and they are clipped off, too. As the peak negative voltages drive the transistor to cutoff, conduction stops. The peak is not passed into the device and cannot appear in the limitier output (Fig. 11-4).

In a discriminator the limiter circuit must be used since the discriminator will detect the AM as well as the FM. The limiter circuit must have a strong signal input. If the signal is weak, even with the device biased to go into saturation and cutoff easily, it won't saturate or cutoff. The peak-to-peak voltage of the signal must be large enough in the limiter input to exceed the saturation and cutoff voltages. If the signal is in the order of a volt or more peak-to-peak the appropriate size is reached.

This limiter threshold voltage is one of the reasons why FM radios are found with more than one stage of IF. Also, the

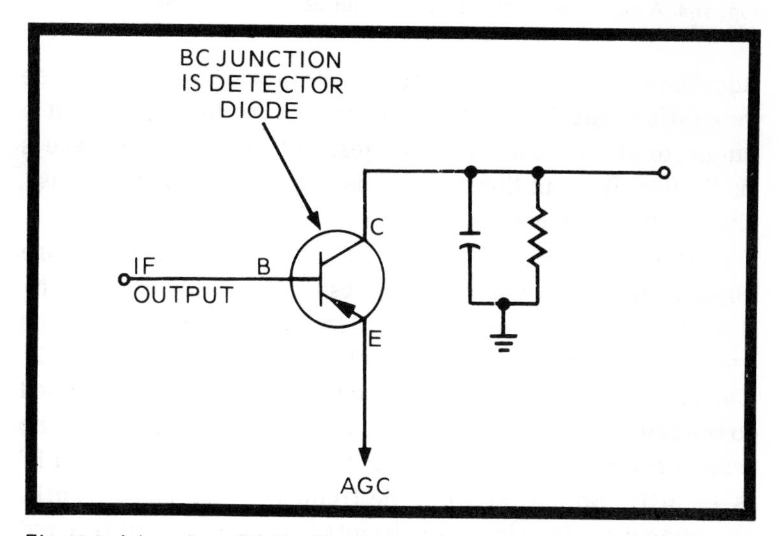

Fig. 11-3. A transistor BC junction also can be used as a detector diode. That way some gain is achieved instead of a loss.

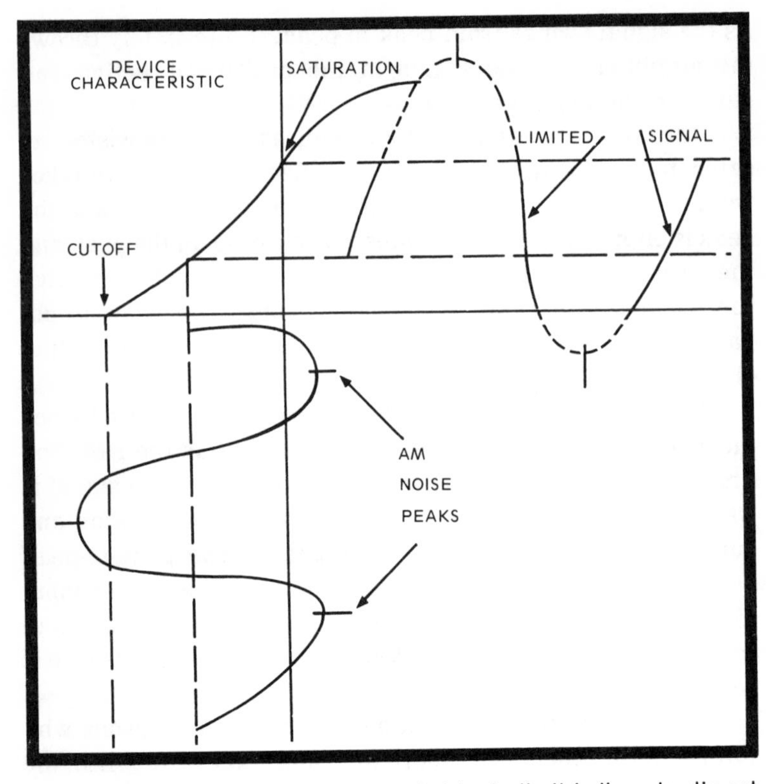

Fig. 11-4. A limiter provides some gain, but basically it is there to clip out AM noise peaks.

additional amplification helps maintain the noise-free reputation that FM enjoys. AM radios are found with but a single, or at the most two, IF stages. And in AM radios there is no limiter. Without this delicate detection that FM radios use, the AM radio is known to be fairly noisy, as a rule.

Once the tops and bottoms of the IF voltage peaks are clipped off in the limiter, the waveshape is ready to enter the discriminator. The discriminator follows the limiter and the two are coupled by a discriminator transformer (Fig. 11-5). The primary and secondary of the transformer are both tuned to the sound IF. In an FM radio the resonant frequency of the transformer is 10.7 MHz and in a TV the intercarrier sound IF is 4.5 MHz. No matter what the frequency, the circuit works the same way. In color sync circuits a similar circuit is found operating at 3.58 MHz.

The output of the limiter is not only fed directly into the primary of the transformer, but a second output branch with a blocking capacitor is attached to the center tap of the secondary. As a result, three distinct voltages are developed in the transformer. Across the primary is V1. Two voltages are developed across the secondary, one from the top of the transformer to the center (V2) and another from the center to the bottom (V3).

Two diodes are attached to the top and bottom of the secondary, with the anodes connected to the transformer. The cathodes of the diodes are connected by two capacitors in series. The capacitor values are chosen to present little or no impedance to the IF. As far as the IF is concerned, the capacitors are ignored and it's as if the cathodes were tied directly together. Two resistors are connected across the two capacitors; they are the diode load resistors with the center tap attached to the secondary center tap. This balances the circuit, placing equal but opposite voltage on the two load resistors.

The idea of the circuit is to develop an output voltage that varies with the frequency modulation. This is accomplished by taking the three voltages from the limiter, rectifying them and adding them together across the load resistor.

Let's consider the actual current flow. At the resonant frequency, the two diodes conduct from cathode to anode. At

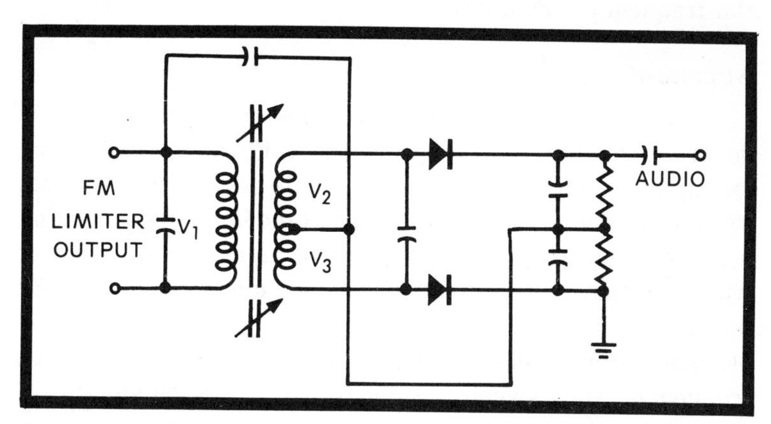

Fig. 11-5. The vertical discriminator circuit extracts modulation from frequency variations.

resonance V2 and V3 are equal, causing equal amounts of conduction. The voltages across the two load resistors are caused by electron flow from the center tap, through the resistors to the diodes. The two voltages across the resistors are equal at resonance. Since they are opposite to each other in polarity, they cancel and the net result of the addition is zero.

When the frequency rises above resonance, an imbalance sets in. V3 becomes more positive than V2. This increases the voltage on the bottom diode anode and decreases the voltage on the top diode anode. There is more conduction in the bottom diode, due to the greater forward bias and the drop across the bottom load resistor is greater. When the voltages across the two resistors add, the net result at the top of the pair is a negative voltage.

When the frequency falls below IF resonance, a similar imbalance occurs. V2 becomes more positive than V3, causing the top diode to conduct more heavily than the bottom. The top diode this time has the greater forward bias, and the voltage drop across the top load resistor is greater. The two voltages add and the net result is a positive voltage. So you can see that a change in frequency in the discriminator input causes a predictable change in voltage in the output. The voltage change is now in amplitude, which the subsequent AF amplifiers can work with. The voltage change is a direct result of the frequency modulation.

Alignment

Discriminator alignment presents some confusion. In general, though, it is a fast procedure if it's understood. Also, an alignment attempt provides valuable service information. The idea is to apply a signal voltage at the input and view it on a scope in the output. The alignment signal is 10.7 MHz or whatever the IF is, with a frequency modulation of 60 Hz usually.

The confusion lies in the modulation. The 60 Hz is a sweep frequency. All a sweep frequency does is modulate the IF around its center frequency. This brings in the mention of another factor that produces more confusion—the width of the sweep. The width or bandwidth of the sweep in a 10.7-MHz IF is usually around 100 kHz. This means that the sweep oscillator is changing its frequency from 10.7 MHz to 10.8 MHz, back through 10.7 MHz and then down to 10.6 MHz. In other words, the sweep oscillator is changing its frequency plus or minus 100 kHz on a continual basis. It is sweeping from side to side around the 10.7-MHz IF frequency. What is the 60-Hz modulation then? The sweep oscillator is sweeping plus or minus 100 kHz 60 times every second.

The scope, of course, cannot ordinarily produce a waveform of 10.7 MHz. Therefore, the scope is attached across the discriminator load resistors. It can reproduce the addition of voltages across the resistors. The voltages will be changing at a 60-Hz rate.

The scope waveform produced by the alignment signal is commonly known as an "S" curve (Fig. 11-6). The "S" curve starts at the base line in the center of the display. When 10.7 MHz is applied to the input, the voltage at the resistors add to zero. As the sweep oscillator goes above 10.7, a negative voltage is developed and the "S" curve drops below and to the right of center. As the sweep goes toward 10.6 MHz, a positive voltage is developed and the "S" curve rises above and to the left of center. If you increase or decrease the width of the sweep, the "S" curve will spread out or narrow.

The actual alignment is simple. Starting with the secondary (bottom core) and then the primary (top core), a good "S" curve on the scope can be produced. The horizontal sweep on the scope is set at 60 Hz, like the sweep oscillator. If you change the scope frequency to 120 Hz, two "S" curves, opposite to each other, will appear (Fig. 11-7). When you get them to cross over at exactly the center in a linear manner, the discriminator is aligned.

RATIO DETECTOR

The ratio detector is a variation of the discriminator and was designed so the limiter stage could be done away with. It contains its own limiting components and provides the discriminator function. It turns out, though, that even though

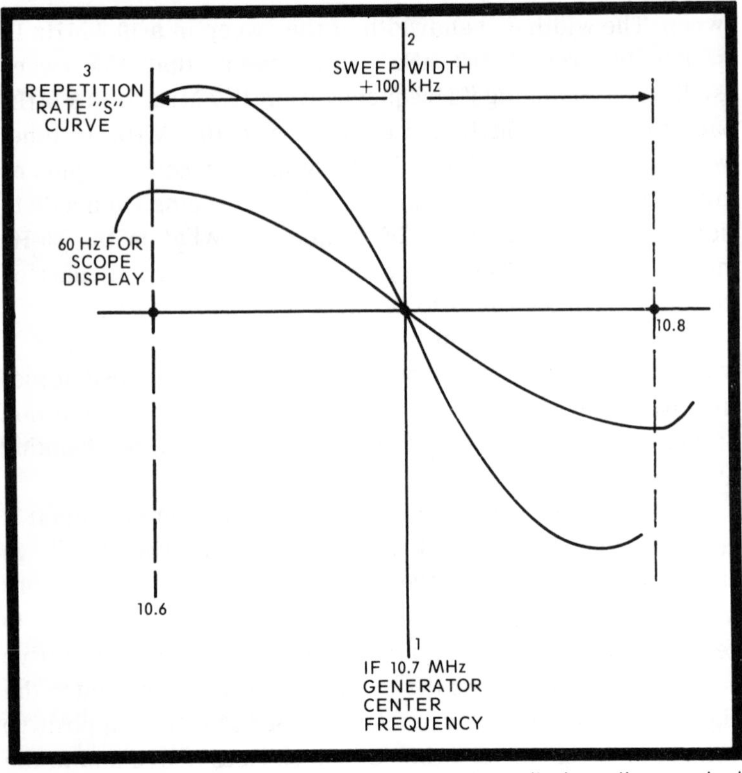

Fig. 11-6. Typical discriminator alignment curve displays three output frequencies.

ratio detectors are put into FM circuitry, quite often limiters are still used to further the noise reduction ability.

At first glance, the circuit looks like the discriminator, but a close appraisal reveals three changes (Fig. 11-8). One, the dual load resistors are replaced with a single load resistor across the two diodes. Two, the bottom diode cathode is attached to the bottom of the transformer secondary. Three, a filter capacitor about 5 or 10 mfd is placed across the single load resistor. As the voltage goes above and below resonance, the capacitors charge according to the amount of conduction through the two diodes. When one diode passes more electrons than the other, its capacitor gets a larger voltage charge. The conduction occurs as the input voltages change along with the frequency change. The total charge across the two capacitors always remains the same, but the ratio between the two

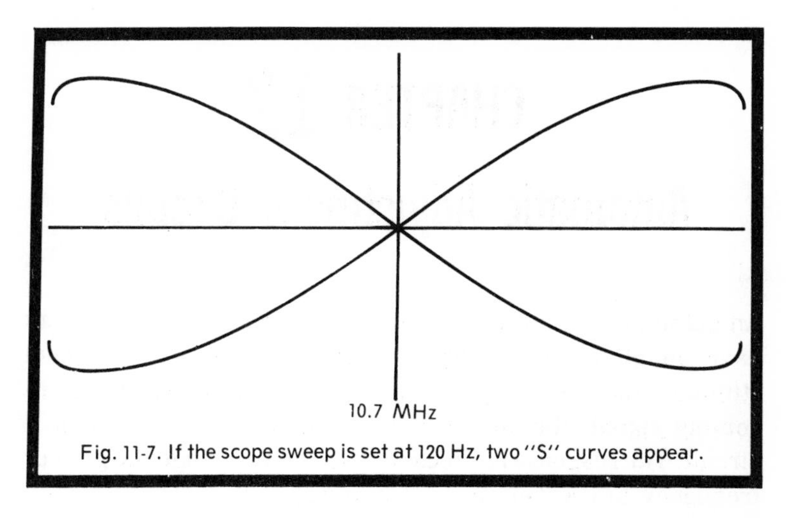

Fig. 11-7. If the scope sweep is set at 120 Hz, two "S" curves appear.

charges changes. At different instants there are different charges across each capacitor. At the center tap between the two capacitors, a voltage difference appears which corresponds to the frequency modulation.

The IF is rectified and causes a current flow through the single large load resistor. The filter across the resistor produces such a high time constant between R and C that not only is all of the IF completely filtered out, but any AM noise bursts are also removed. In effect, this RC network is the limiter.

Visual alignment is best, but follow the manufacturer's service notes for variations. The RC network usually has to be removed for alignment and a bias voltage applied in its place.

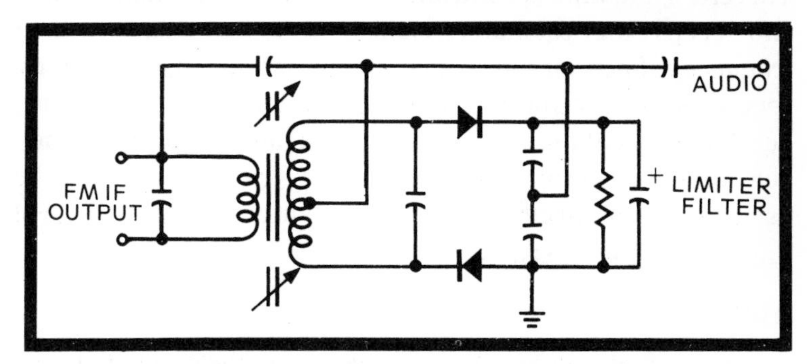

Fig. 11-8. The ratio detector is similar to the discriminator circuit. The audio is developed by the voltage ratio across the series capacitors.

CHAPTER 12
Automatic Adjustment Circuits

An automatic circuit offers a certain fascination. It seems to have an almost living quality since such circuits receive stimulus and then respond in turn. The stimulus is the incoming signal, the output of an oscillator or the gain in a circuit. The response is a feedback that can change the gain or frequency to compensate in some way.

Most automatic circuits break down into two types. The first is a circuit that takes a sampling of gain and then feeds back a control voltage to adjust the amplification so the final signal output remains constant rather than vary in strength as the incoming signal does (Fig. 12-1). This type of circuit is aptly called (AGC) automatic gain control, or in the case of a radio it is called automatic volume control (AVC); both are identical.

The other circuit takes a sampling of an oscillator frequency, compares it to another frequency, and produces a DC correction voltage. The correction voltage is fed to an oscillator directly, or is fed to another circuit that forms a capacitance or inductance in the oscillator tank circuit. Therefore, the tank is controlled by the DC correction voltage and keeps the oscillator locked in to a predetermined frequency. Such circuits vary widely and are used in many receivers to control the local oscillator, color oscillator, horizontal oscillator. They are called automatic frequency control (AFC) circuits (Fig. 12-2). AFC and AGC can both be used to control circuits that are designed to control several functions simultaneously.

AGC

An AGC circuit has the job of reducing the amount of gain in a given circuit, but it cannot increase the gain of a circuit

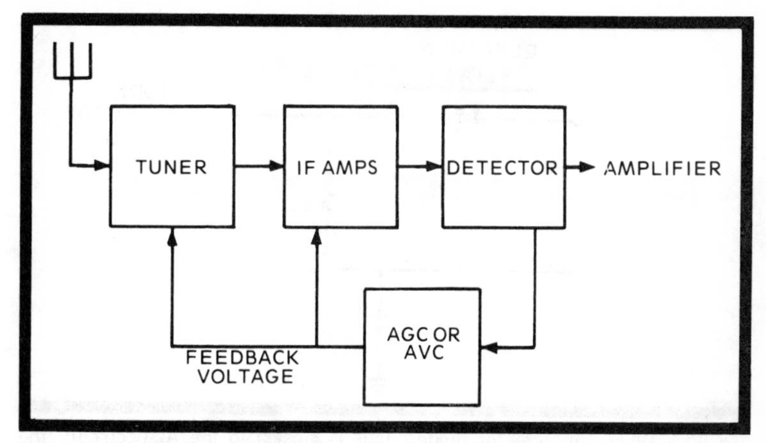

Fig. 12-1. The AGC circuit takes a portion of the detector output and feeds it to the receiver input to adjust gain.

beyond its design limitations. The AGC circuit derives its voltage from the detector circuits discussed in Chapter 11.

The simplest AGC circuit is nothing more than a resistor and capacitor attached to the output of a detector circuit (Fig. 12-3). The time constant of the two components has to be large enough to filter out the detected signal, but the time constant can't be too large. If it is too large, the AGC doesn't respond quickly enough to changes in the peak-to-peak level of the

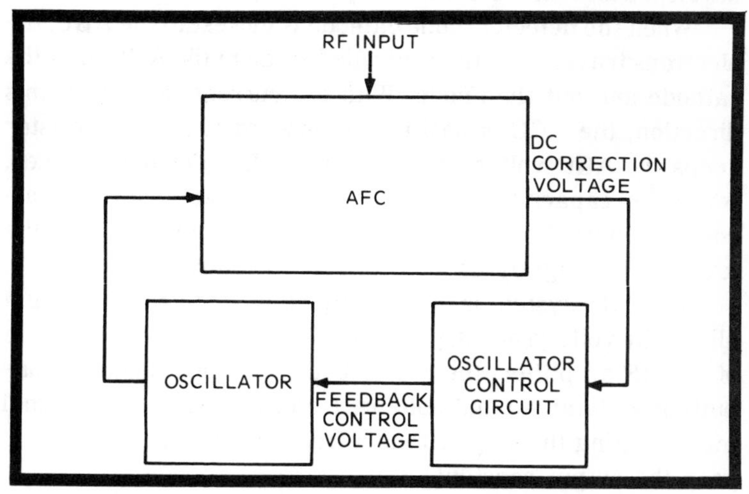

Fig. 12-2. The AFC circuit takes the RF input and compares it with the oscillator frequency. A correction voltage is produced to synchronize the oscillator with the RF.

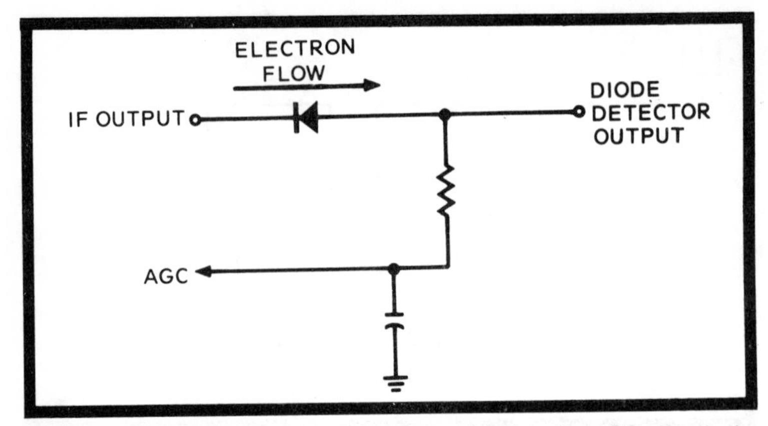

Fig. 12-3. When the detector diode anode is closest to the AGC circuit, the output is a negative voltage.

incoming signal. A typical AM radio AGC time constant is a tenth of a second. It is obtained with a resistor in the megohm range and a capacitor around .05 mfd.

Here's how an AGC circuit works: The AM signal is rectified and detected. If the detector diode anode is closest to the AGC, the rectified signal current travels toward the AGC, through the megohm isolation resistor and then on to the stages that need control (Fig. 12-4). A negative charge is developed and the AGC is said to provide reverse bias.

When the detector diode cathode is closest to the AGC, the electrons travel from the controlled stage to the AGC, into the cathode and out the anode. With the current flowing in this direction, the AGC is positive or forward bias. The resistor keeps the AGC voltage isolated from the controlled stages, while the capacitor keeps the DC at a steady level in accordance with the incoming signal strength, even though the level varies slightly as it performs its control.

The AGC system is more complex in some receivers, but all use the voltage developed at the detector to control the gain of the RF amplifier, the IFs or both. A strong signal means putting a clamp on and holding it down while a weak signal means letting the stage run in a freer fashion. The point is to keep the output constant.

The AGC system in some TV receivers employs three stages. One is usually called the AGC gate; two, the AGC

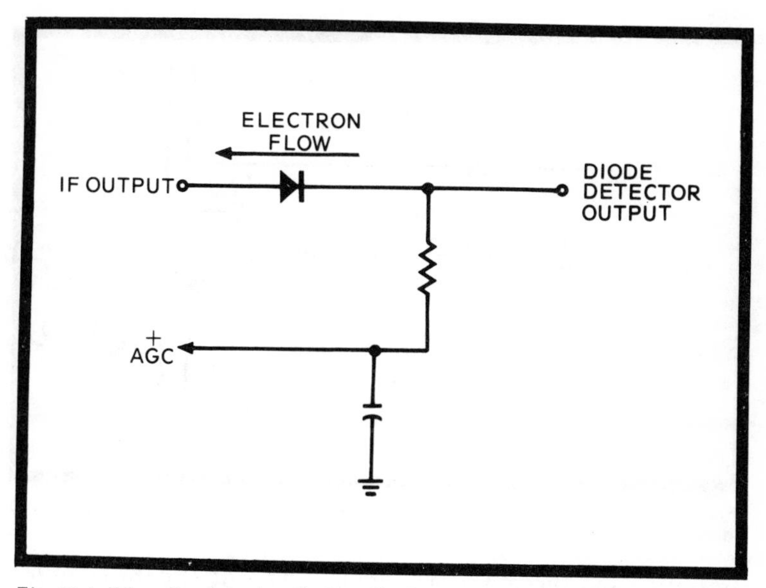

Fig. 12-4. When the detector diode cathode is closest to the AGC circuit, the output is a positive voltage.

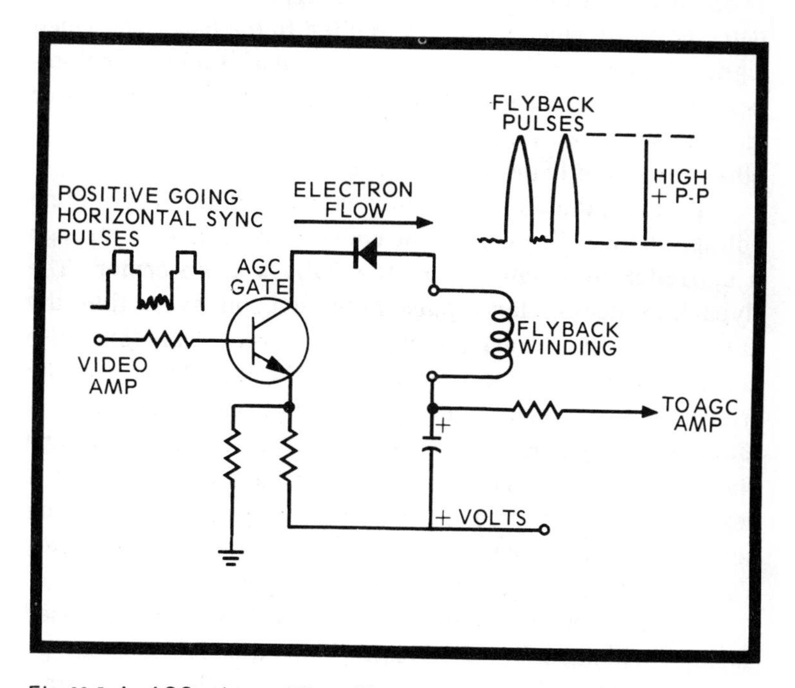

Fig. 12-5. An AGC gate amplifier will not conduct unless the flyback pulse and sync pulses are both present.

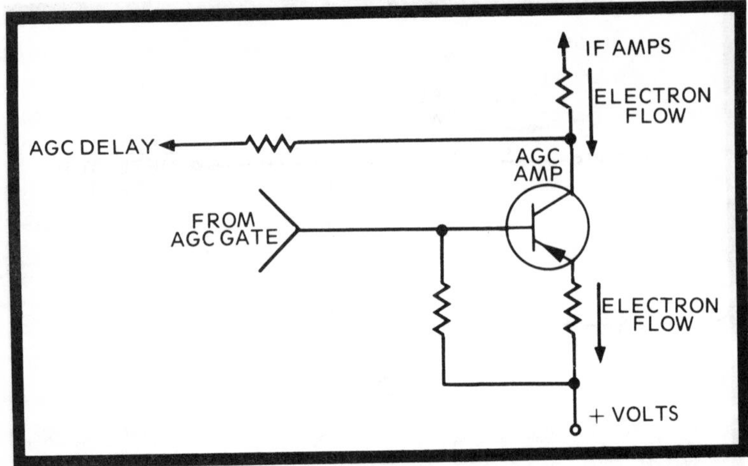

Fig. 12-6. An AGC amplifier makes the varying AGC DC voltage into a much larger voltage.

amplifier and three, the AGC delay. The AGC gate is called that since it will not operate unless it is keyed or gated (Fig. 12-5). In a TV, an incoming signal sample is derived from the detector or the video amplifier and fed to the base. The video signal also contains a horizontal sync pulse at a high peak-to-peak level. The input gate is biased so that the transistor cannot conduct until the sync pulse arrives at the base. At all other times the transistor is cut off.

However, that is not the only cutoff medium. There is no voltage on the collector, but the collector is connected through a capacitor to a winding on the flyback transformer. The flyback produces a large pulse in the winding every time the horizontal output transistor switches on and off and this pulse is applied to the collector. The pulse on the collector causes the transistor to conduct every time the flyback pulse appears. Simultaneously, the sync pulse on the base should forward bias the transistor. The transistor conducts every time it is keyed at its base and collector by the two pulses. This keying action eliminates noise pulses that appear atop the video. Thus, the AGC is a clean pulsating output.

In the collector circuit is a diode that is forward biased during each pulse. The diode is there so the electrons can go through it but can't possibly go back to the collector and start the stage oscillating, or ringing as it's called.

Next, the electrons encounter the negative side of a large filter capacitor. An isolation resistor passes the electrons that are discharged from the filter to the next stage, the AGC amplifier. The filter removes the 15,750-Hz horizontal pulse, and supplies a steady DC. The charge on the filter is an excess of electrons so it is a negative charge. This negative charge is applied to the AGC amplifier (Fig. 12-6). It is forward biased, since the AGC amplifier is a PNP. PNPs have negative bias on the base and positive on the emitter.

The AGC amplifier conducts in exact accord with the AGC variations. A small variation in the base of the amplifier becomes a large variation in the collector current. In the PNP the collector current is flowing from the collector to the emitter. In other words, the collector is taking electrons from the AGC line leading to the IF stages and also from the emitter of the AGC delay circuit. The electrons from the IF are taken from the bases of the first and second IFs. Since they are both NPNs, taking electrons from the circuit increases the forward bias on the two IFs, thus controlling the conduction.

By taking electrons from the RF AGC delay's emitter (Fig. 12-7), its forward bias is increased. The AGC delay is so called since it does not turn on as soon as the IF AGC. This is because the RF transistor is best run wide open. The high RF gain fed to the mixer stage prevents excessive conversion loss and subsequent snow in the TV picture. The RF transistor is

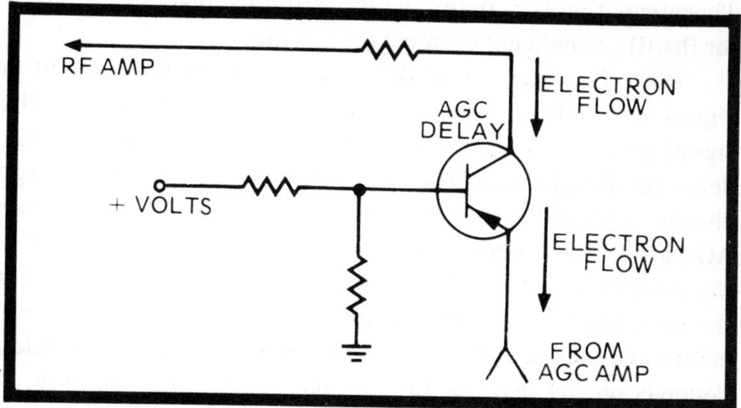

Fig. 12-7. The AGC delay circuit is between the RF amplifier and the AGC. It operates only during high signal levels.

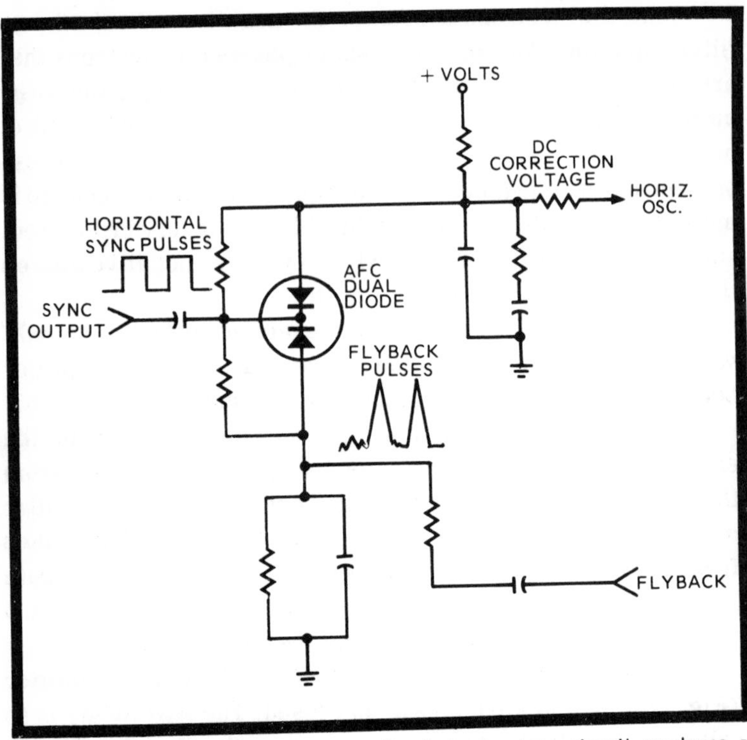

Fig. 12-8. The AFC dual diode in a discriminator type circuit produces a correction voltage from the amplitude of the horizontal sync pulse.

designed not to receive any AGC voltage until so much signal is coming in that the RF transistor is about to overload. Therefore, the AGC delay, which is the exclusive AGC agent for the RF, is held cut off most of the time.

There is a significant difference between the amount of signal needed to turn on the IF AGC and the RF AGC. As the signal gets so strong that the RF tends to overload, the AGC delay finally turns on. The sync tips get so large that a large charge is stored in the AGC filter. This forward biases the AGC amplifier to draw so many electrons from the emitter of the delay transistor that it turns on. Electrons are drawn out of the RF transistor's base so that the NPN transistor in the RF is forward biased and a gain reduction takes place. The delay depends on the level of the voltage rather than on time. It takes a higher level of input voltage to reduce the RF transistor gain than it takes to reduce the IF transistor gain.

AFC

Automatic frequency control is different from automatic gain control. AFC utilizes feedback to keep an oscillator under control, such as the local oscillator, the horizontal oscillator and the color oscillator (the vertical sweep oscillator is not mentioned because its 60-Hz frequency can be controlled directly from the sync signal in the composite TV signal).

The horizontal oscillator is controlled with an AFC circuit (Fig. 12-8). Many of them are identical to the discriminator described in FM detection (see Chapter 11). The only difference is that one input is the horizontal sync pulse in the TV signal. A second input to the discriminator is taken from the horizontal output transformer. The two input signal frequencies are compared and their difference comes out of the discriminator as a DC correction voltage. This voltage sets the bias in the horizontal oscillator. As the oscillator attempts to drift off frequency, the spike from the flyback changes and causes the DC correction voltage to vary, bringing the oscillator back to its designated frequency. The DC correction voltage is produced exactly like the AF output in an FM radio.

This type of AFC is common in color circuits, too (Fig. 12-9). The color oscillator is supposed to run at exactly 3.58 MHz.

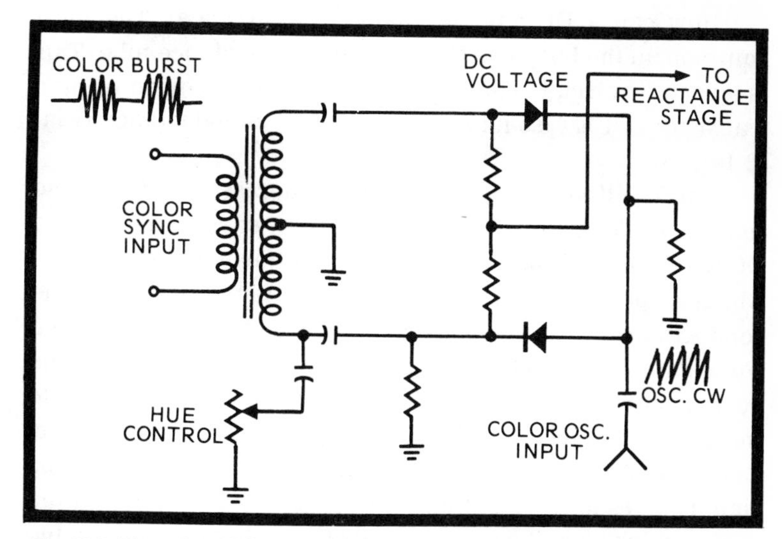

Fig. 12-9. The color AFC circuit produces a DC correction voltage to sync the color oscillator with the color burst.

Some of the oscillator energy is tapped off and inserted into the color AFC discriminator. Also injected into the phase detector is the color sync signal—the burst—from the composite TV signal. The burst and the oscillator signal frequencies are compared and any difference results in a correction voltage.

The CW from the oscillator is sent in through the output. The burst is coupled into the stage through a coupling transformer. The DC correction voltage is taken off either at the output or the center tap between the load resistors. Various configurations exist, but they all are basically a discriminator.

The DC correction voltage from an AFC stage can be the result of a single input or the comparison of two inputs. The single input is the transmitted signal, just like in an AGC system, except it is fed to the base or gate of the local oscillator to keep it locked in step. The DC correction voltage can control an oscillator frequency by varying the bias on the oscillator. However, there is a second method of oscillator lock in; that is, using the DC correction voltage to vary an inductance (measured in microhenrys) or a capacitance (measured in picofarads).

How can a DC voltage change a reactance? The most common method through the years is the reactance tube. This is a specific circuit. The circuit can also be designed using a transistor or FET, so let's use an FET as an inductance (Fig. 12-10).

The oscillator tank is connected across the drain and source. A blocking capacitor keeps the two circuits isolated DC wise. The DC correction voltage is attached to the gate to which is applied the correct control bias. As the DC varies, the conduction through the FET varies. A capacitor is installed from gate to ground and a resistor and capacitor are connected between the gate and drain with the resistor closest to the gate. The oscillator signal has to pass through the stage, since the stage is part of the oscillator tank. The oscillator signal comes in through the blocking capacitor, arrives at the drain, which it can't enter, and goes through the capacitor and resistor between the drain and gate. The gate-to-ground capacitor begins to charge and discharge in step with the frequency. The current at the gate input is in phase with the oscillator.

It is a fact about capacitors that the voltage across it lags the current by about 90 degrees. That's because it takes about 90 degrees of time for the capacitor to charge from the incoming current. The current arrives first and the voltage comes up to value 90 degrees later when the capacitor fully charges. It is also a fact that the drain current of an FET is in phase with the gate voltage. Since the gate voltage is 90 degrees behind the gate current, the drain current is likewise 90 degrees behind the gate current. The conclusion of the sequence is that an oscillator voltage applied to this circuit causes the FET to produce an output current 90 degrees behind the oscillator voltage.

An inductance is a device that introduces a 90-degree lag between the current and voltage. Therefore, the FET and its components have become an inductance. Furthermore, the amount of DC correction bias on the gate varies the amount of inductance. Since the inductance controls the tank circuit and, therefore, the oscillator frequency, an AFC condition is created. A more positive control voltage, which further turns on the FET, reduces the inductance in the circuit and a more negative DC correction voltage pinches off some of the FET conduction and produces a smaller amount of inductance.

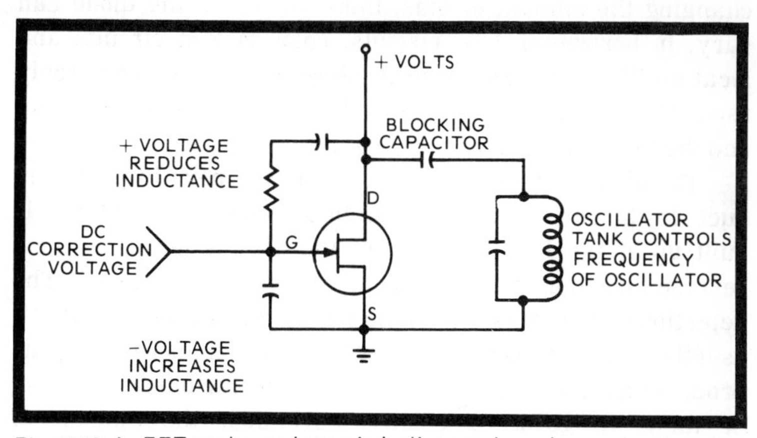

Fig. 12-10. An FET can be used as an inductive reactance by causing its output current to lag its output voltage by about 90 degrees.

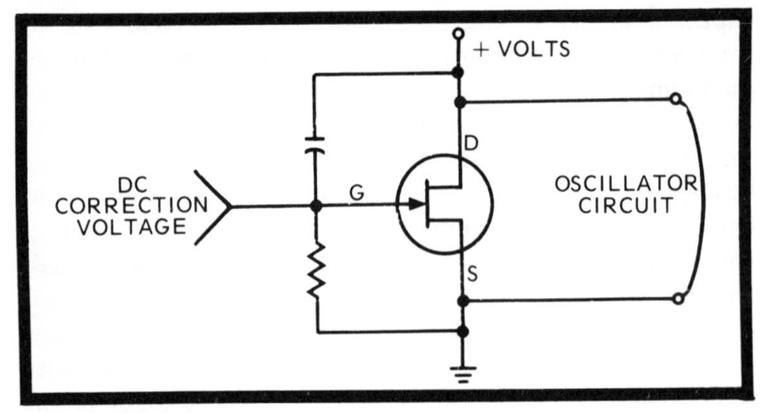

Fig. 12-11. An FET can be used as a capacitive reactance by causing its output voltage to lag its output current by about 90 degrees.

A capacitive reactance can be produced by simply reversing the resistor and the gate-to-ground capacitor in the circuit (Fig. 12-11). This makes the voltage in the circuit lag the current by 90 degrees, which is what a capacitive reactance causes.

VARACTOR DIODES (VARICAPS)

While the reactance amplifier circuit is useful in solid-state circuits, the simplest way to achieve AFC is with varactor diodes. As mentioned in the beginning of the book, a diode has the ability to act just like a capacitor. Also, by changing the amount of bias, the capacity in the diode can vary. In horizontal AFC circuits, color lock-in circuits and local oscillator control circuits, these diodes are commonly used. The circuitry is much simpler than a reactance circuit and they are less expensive to produce and install.

The diode is reverse biased and never permitted to conduct. In the reverse bias mode, an amount of capacitance is built up at the PN junction. As the bias is increased in a reverse direction, the amount of capacity is increased. The depletion of electrons across the junction is increased and it's as if the surface area of the dielectric has been increased. This produces a larger capacitance. As the reverse bias is reduced, or made to go in a positive direction, the amount of capacitance is decreased as the dielectric is made effectively smaller. It follows that if a DC correction voltage is varying the reverse bias, it can control the amount of capacitance. If the capacitance is installed in a tank circuit or other frequency controlling role, it can lock in an oscillator.

Typically, a diode is placed in a TV tuner directly across the oscillator coil (Fig. 12-12). Two DC isolation low-value capacitors tie the diode to the coil. Two resistors form a voltage divider on either side of the diode to set the correct amount of reverse bias. The cathode of the diode is attached to the voltage end and the anode to the ground end. A positive voltage applied to the cathode reverse biases the varactor. The two tiny capacitors prevent DC from getting to the oscillator coil but the capacitances are so small that they do not enter into the tank circuit operation in any appreciable manner.

The DC correction voltage is applied along with the DC bias. It comes from the output of a special discriminator circuit. The uniqueness is due to a DC transistor amplifier to amplify the DC correction voltage and a zener diode to keep the amount of DC bias at a rock steady level. Any DC bias variations are coupled into the varactor and cause erratic operation of the local oscillator. See the following charts for AGC and AFC circuit tests.

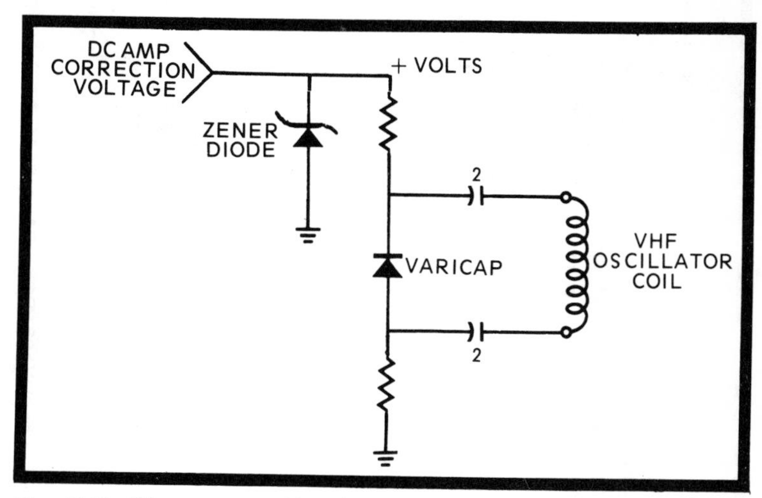

Fig. 12-12. When reverse biased, a varicap diode becomes a variable capacitance by changing the amount of reverse bias.

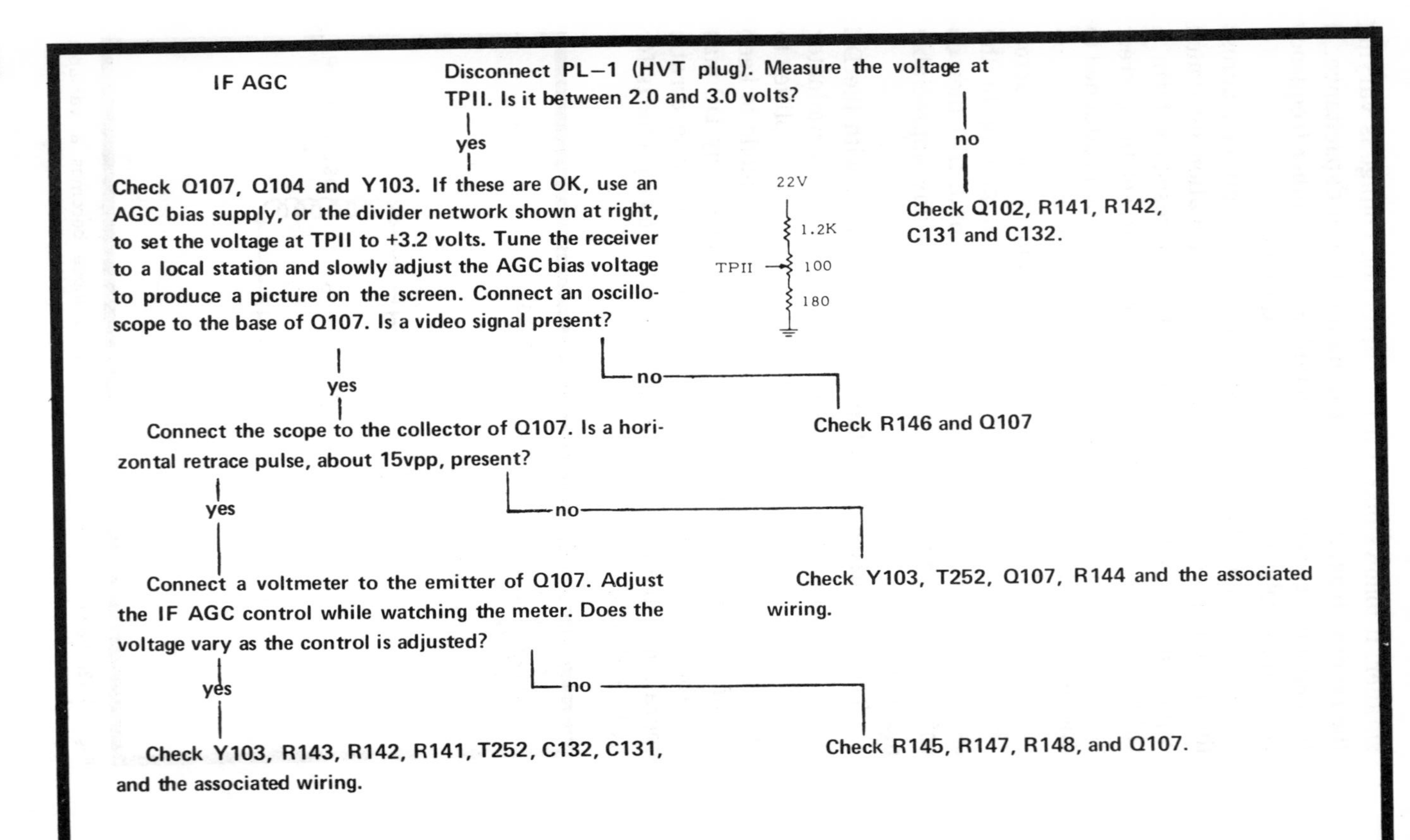
IF AGC
Disconnect PL—1 (HVT plug). Measure the voltage at TPII. Is it between 2.0 and 3.0 volts?
yes
no
Check Q107, Q104 and Y103. If these are OK, use an AGC bias supply, or the divider network shown at right, to set the voltage at TPII to +3.2 volts. Tune the receiver to a local station and slowly adjust the AGC bias voltage to produce a picture on the screen. Connect an oscilloscope to the base of Q107. Is a video signal present?
22V
1.2K
TPII
100
180
Check Q102, R141, R142, C131 and C132.
yes
no
Connect the scope to the collector of Q107. Is a horizontal retrace pulse, about 15vpp, present?
Check R146 and Q107
yes
no
Connect a voltmeter to the emitter of Q107. Adjust the IF AGC control while watching the meter. Does the voltage vary as the control is adjusted?
Check Y103, T252, Q107, R144 and the associated wiring.
yes
no
Check Y103, R143, R142, R141, T252, C132, C131, and the associated wiring.
Check R145, R147, R148, and Q107.

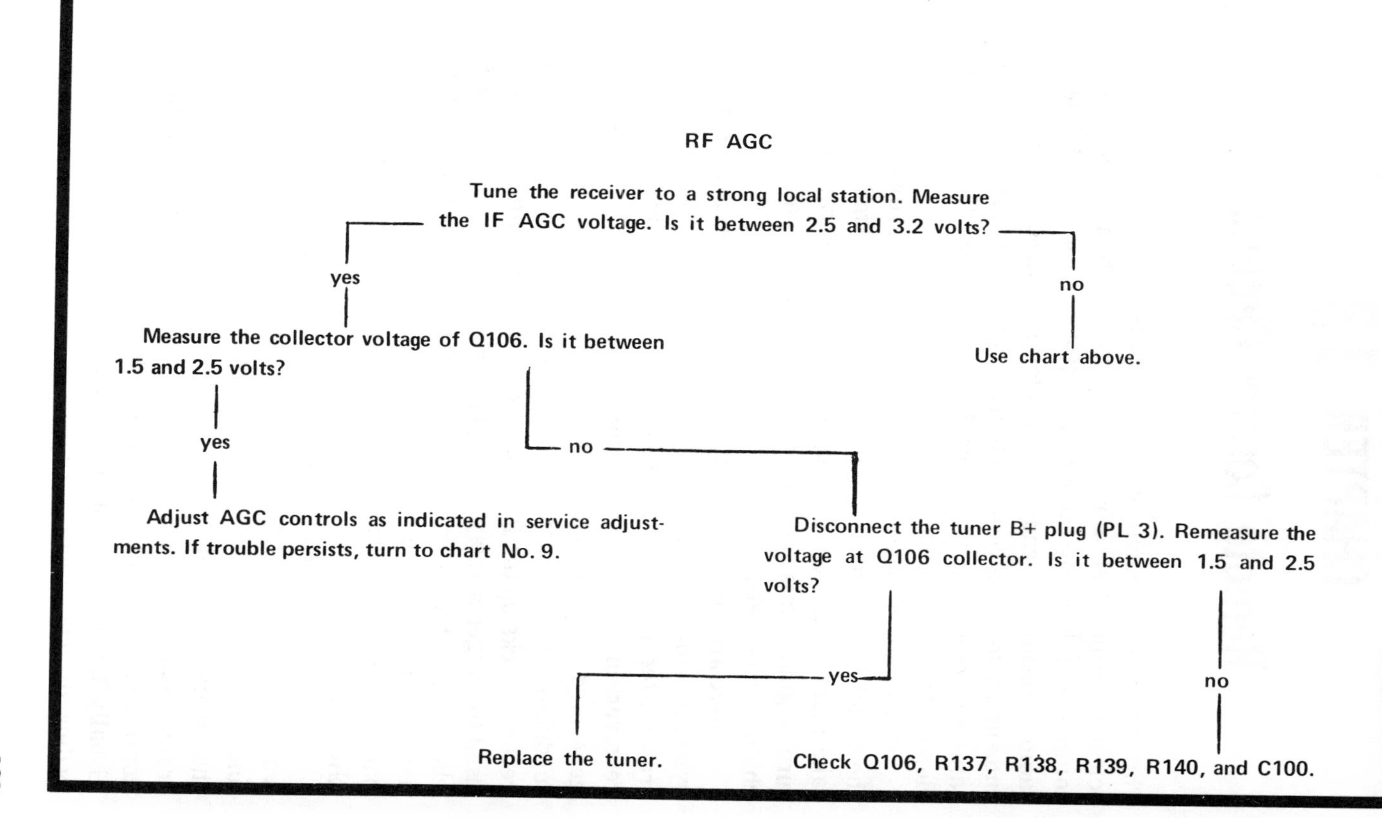
RF AGC
Tune the receiver to a strong local station. Measure the IF AGC voltage. Is it between 2.5 and 3.2 volts?
yes
no
Use chart above.
Measure the collector voltage of Q106. Is it between 1.5 and 2.5 volts?
yes
Adjust AGC controls as indicated in service adjustments. If trouble persists, turn to chart No. 9.
no
Disconnect the tuner B+ plug (PL 3). Remeasure the voltage at Q106 collector. Is it between 1.5 and 2.5 volts?
yes
Replace the tuner.
no
Check Q106, R137, R138, R139, R140, and C100.

CHAPTER 13
Remote Control Circuits

The mention of remote control circuits evokes thoughts of science fiction. Yet a remote control system is no more complicated than any radio transmitter and receiver. While a radio transmitter sends out an electromagnetic wave that ends up as sound, a remote control transmitter sends out the same kind of wave that ends up flicking a switch. That's all the electronic section does.

Of course, there is a lot more to the actual hardware. Once the switch is actuated, a motor turns on and all kinds of events take place. A garage door opens, a drone aircraft makes a turn, a guided missile rises from a launching pad, a TV set changes channels, etc.

There are two common types of electronic remote control systems. One is a long-range operation, which means the transmitter is separated a considerable distance from the receiver. In remote applications of this nature, an RF signal is generated in the transmitter. The encoded information modulates the carrier wave and then is transmitted. The receiver picks up the RF and detects the modulation, which is used to trigger a relay that opens or closes a switch or switches. That's the extent of the electronics. Transistors, FETs and ICs are very useful in these applications. The compactness and low weight is of great value in lots of applications.

The second type of remote system uses an ultrasonic air wave, rather than RF. This type of system is much simpler and more reliable than the RF. However, its big limitation is that it can be used only in applications that have a short transmission distance. A typical application is the remote control of a TV set, from one side of a room to another. Usually, this is the extent of the distance. It can't even be used ordinarily from room to room.

RF REMOTES

RF remotes are commonly used in model airplane work, garage door openers, etc., as well as the highly sophisticated military and space age applications (Fig. 13-1).

The simplest remote system employs a device similar to a telegrapher's key as the modulator. The RF is produced and transmitted. A button or a key is depressed a number of times. In the receiver, there may be three relays. One dash actuates relay number one. Two dashes actuates number two and three dashes actuates the third relay. The button simply turns the RF on and off and the receiver is tuned to the interruptions. This is a common way to control simple model airplanes.

Other models of control employ a transmitter that can send out a number of different radio frequencies. One channel turns on one motor and a different channel turns on a second motor. Or one frequency could be modulated with different tones like 400 Hz, 1000 Hz and then 2000 Hz. The encoding actually can become complex, but basically that's about all there is to the idea of it.

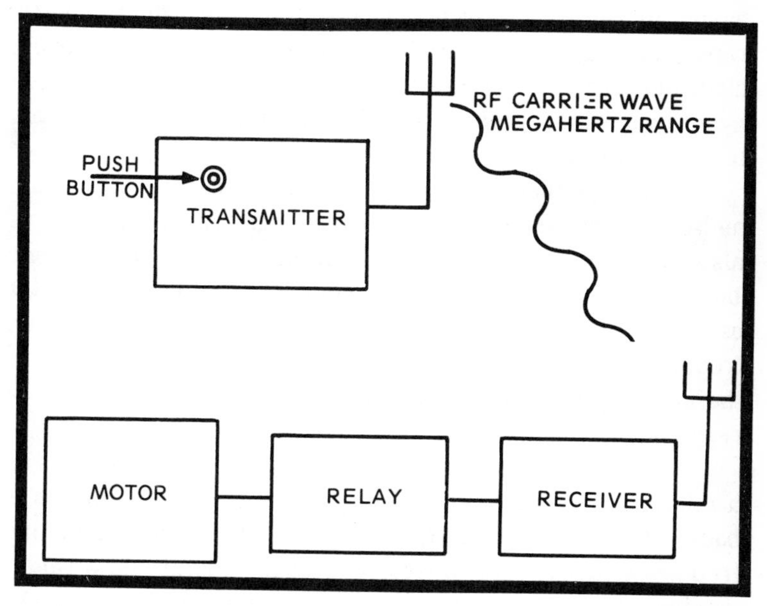

Fig. 13-1. When a button is pushed in a remote transmitter, a relay is activated in the remote receiver.

The receiver has an antenna that is tuned to receive the transmitted frequencies. There is an RF amplifier, perhaps a converter, IF stages and a detector. The system could be AM or FM. After the detector, however, there usually are no audio stages (unless an audio signal monitor is needed). Instead of the audio, the detected information, which could be a sine wave tone, square-wave pulses from on-off RF, or other types of modulation, is sent to a relay control transistor.

The relay control is a DC amplifier. Since the detected signal is a small voltage, it is sent into a heavy-duty transistor and the signal turns on the relay transistor. Electrons flow heavily between emitter and collector and the current activates the relay. The relay opens or closes a switch and the mechanism goes to work. An operator at a great distance can cause work to be performed.

TRANSDUCER REMOTES

When the remote function must cover only a short distance, a high-frequency signal is not needed. It is much easier to simply produce a frequency in the 30- to 50-kHz range, just above the audible frequencies. At this frequency, electromagnetic waves are not needed; an air wave, produced by a tiny speaker, passes from the transmitter to the receiver to do the work.

A tiny speaker that can efficiently generate these non-audible air waves is called a transducer. The transducer is also an efficient microphone, a capacitor-type microphone built to reproduce the desired frequency. A simple transistor oscillator can produce a number of frequencies. Each frequency is passed into the transducer, converter to energy in the air, then passed through the air from the transmitter to receiver.

In the receiver is another identical transducer that acts as a microphone. The mike picks up the signal and converts it back into electrical energy. A trap tuned to the particular frequency that is transmitted couples the signal into a relay-driving transistor. The relay is actuated and the system performs the function.

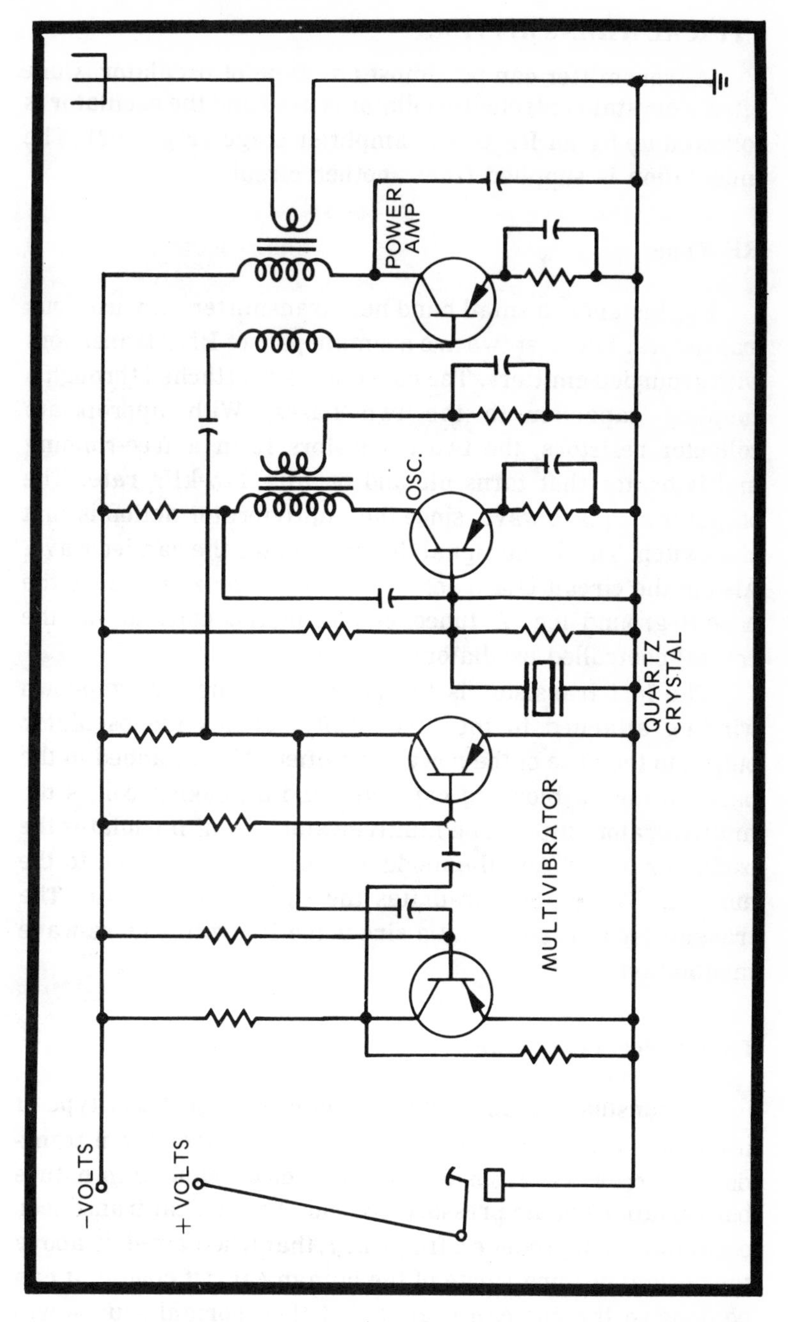

Fig. 13-2. This remote transmitter uses a multivibrator to modulate a crystal oscillator. (Courtesy Texas Instruments)

TYPICAL TRANSMITTERS

A transmitter can be almost any type of oscillator. Quite often a crystal-controlled oscillator is used and the oscillator is followed up by an RF power amplifier stage (Fig. 13-2). The modulation is supplied from another circuit.

RF Type

For instance, a small hand-held transmitter may use four transistors. The first two are a simple pair of PNP transistors with grounded emitters. The collectors are attached through a coupling capacitor to the two bases. With appropriate collector resistors, the two transistors form a free-running multivibrator that turns off and on at a low-kHz rate. The output is a square wave, since the multivibrator action is that of a switch. This is the signal that modulates the carrier wave. Also in the circuit is a transistor with a quartz crystal in the base-to-ground leg. A tuned coil is in the collector of the crystal-controlled oscillator.

The last transistor is the power amplifier. A stepdown winding attached to the tuned coil matches the oscillator output to the base of the power amplifier. Also attached to the base through a blocking capacitor and a peaking coil is the multivibrator output. The multivibrator output modulates the oscillator output and the modulated signal is matched to the antenna. The antenna radiates the signal into the air. The transmitter thus sends out a sine wave RF with square-wave modulation.

Transducer Type

A transducer-type transmitter cannot be just any type of oscillator. Generally, modulation cannot be added to a transducer carrier wave, since it is not electromagnetic in nature but variations of air pressure instead. The typical transducer oscillator must produce a frequency, that when aired, is above the normal hearing range of the human ear. Of course, it can be done so the ear can hear it, but then normal sounds will interfere with the transmission and reception of the signal.

Unfortunately, there are lots of TVs with remotes that are not selective enough. Channels change or other functions are actuated for no reason at all.

Most systems operate in the frequency range from 35 to 45 kHz. These frequencies can be produced with an oscillator that is running at exactly one half of the transmitted frequency, because a transducer, due to its physical structure, doubles any frequency as it converts it from electrical to audible energy. The doubling occurs because the transducer works as a capacitive speaker or microphone.

There are two plates to the transducer just like any capacitor. As a voltage potential builds up from the oscillator frequency, the plates are attracted to one another. They are pulled together on the positive half of the cycle. Air is squeezed from between the metallized plates. Then, as the cycle hits the zero base line, the plates return to the rest position. As the cycle then goes into its negative region, the plates are again pulled together, forcing air out from between them. For each cycle of signal, the plates are attracted to each other twice, once during the positive peak and once during the negative peak. This doubles the oscillator frequency.

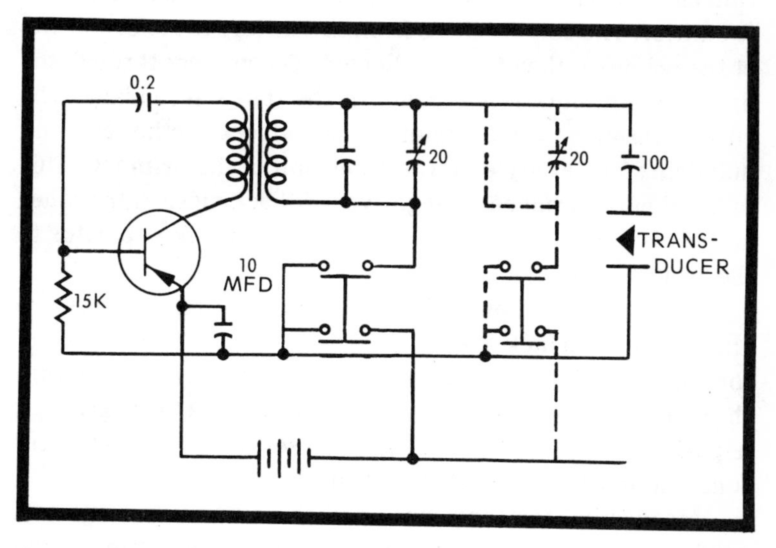

Fig. 13-3. A supersonic transmitter produces an air wave. (Courtesy Magnavox)

A typical oscillator circuit employs a PNP transistor with a tuned transformer in its collector and feedback from collector to base from the bottom of the primary of the transformer (Fig. 13-3). In the Magnavox color TV remote control transmitter, the circuit works as follows:

The collector is coupled to the transducer through a tuned transformer and a small coupling capacitor. While there is only a single transformer, there are a number of variable capacitors that can be switched in across the secondary of the transformer to produce the desired frequency. Each tuned capacitor adds a bit more capacity to the tuned circuit and thus makes a higher frequency available.

The other end of the transducer is returned to the base of the oscillator through a base leak resistor. The emitter is bypassed heavily with a filter capacitor to the bottom of the transducer, too. The emitter is also attached to the battery supply. Each switch is capable of closing the circuit and each adds another capacitor to the tuned transformer secondary.

When a button is pressed, the battery puts 9 volts DC across the PNP transistor. The collector gets a negative voltage, which means there is an excess of electrons on the collector. Electrons flow from collector to base and collector to emitter. The current flows from the battery to the center tap of the primary, then to the collector. The current through the half winding induces a current in the tuned secondary. The current through the half winding also induces another current, auto-transformer style, in the other half of the primary. This induced pulse produces a negative feedback pulse in the other half. This pulse is coupled through the feedback capacitor to the base.

Base current flows from base to emitter. A positive charge, due to a lack of electrons, develops on the base side of the capacitor. Then the pulse passes. The positive charge from the capacitor attracts all the electrons from the base, giving it a positive charge. This reverse biases the base-to-emitter junction and the transistor cuts off.

Meanwhile, the tank circuit in the secondary rings and sends another negative pulse from the transformer to the base. The transistor turns on and the oscillator operates. The signal across the tank is designed to have a very high peak-to-peak voltage. This voltage is transformed into physical energy in the transducer. The more powerful it is, the stronger the two plates will be attracted to each other in the transducer.

TYPICAL RECEIVERS

The typical remote receiver is quite like any radio receiver, except for the fact that instead of an output of a communicative nature, it has an output that actuates switches and relays for some type of motorized work. A receiver has an RF antenna and subsequent RF and detection circuits in the RF type remote. If the receiver is receiving a supersonic air wave, its antenna is a transducer identical to the transducer in the transmitter.

The antenna receives the RF signal, at microvolt level. The supersonic transducer is mounted in an opening in the receiver and its plates are moved by the variations of air pressure caused by the companion transducer in the transmitter. There is a DC bias in the small microphone, and when the plates move, the capacitive change causes modulation in the DC bias. This modulation is coupled through a capacitor to an appropriate staging of amplifiers.

In the RF type receiver, the amplifiers handle a MHz range of frequencies as described in Chapters 5 and 6. In the transducer type, the amplifiers handle a low-kHz range of frequencies as described in Chapter 7. Then the RF signal is detected in similar fashion as the circuits in Chapter 11. In a transducer receiver, the particular frequency processed is picked out by a tuned circuit. There are a number of tuned circuits, one for each of the designed frequencies.

RELAY DRIVERS

Finally, the signal is ready to perform some work. A typical circuit is the Magnavox color TV remote (Fig. 13-4). A tuned circuit receives a signal. The tank is in the primary of a coupling transformer. The signal is passed to the stepdown secondary. Then it is coupled into the base of an NPN relay-

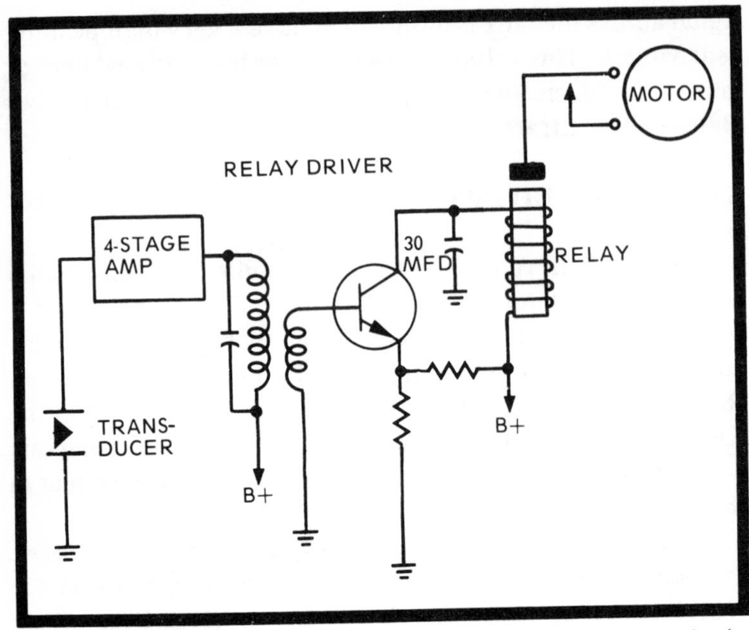

Fig. 13-4. A supersonic receiver picks up the transmitted air wave also by means of a transducer.

driver transistor. A stepdown transformer matches the signal to the base of the transistor. The emitter has a resistor to ground and some DC comes through a large value resistor to the emitter resistor, placing a small positive voltage on the emitter. This reverse bias cuts off the electron flow from emitter to collector.

In the collector leg there is a large value filter and a relay coil to B+. With no conduction, the relay coil has a bias in it, but no magnetic field since the current is not moving. Then the signal from the transmitter arrives at the base of the transistor. The positive-going signal biases the base into conduction. Current flows heavily from emitter to collector as the base switches on the transistor. Current flows through the relay coil and the collector filter charges. Then the negative part of the signal arrives. The transistor cuts off. However, the filter takes over and discharges through the relay coil to the B+ supply. That way, as long as the button is held down in the transmitter, current keeps flowing in the relay coil, first from transistor conduction and then from the filter discharge.

As the current flows in the relay coil, the contacts are attracted by the induced magnetic field and held in a closed position. The contacts are the off-on switch for the motor. The motor keeps running as long as the contacts are held closed.

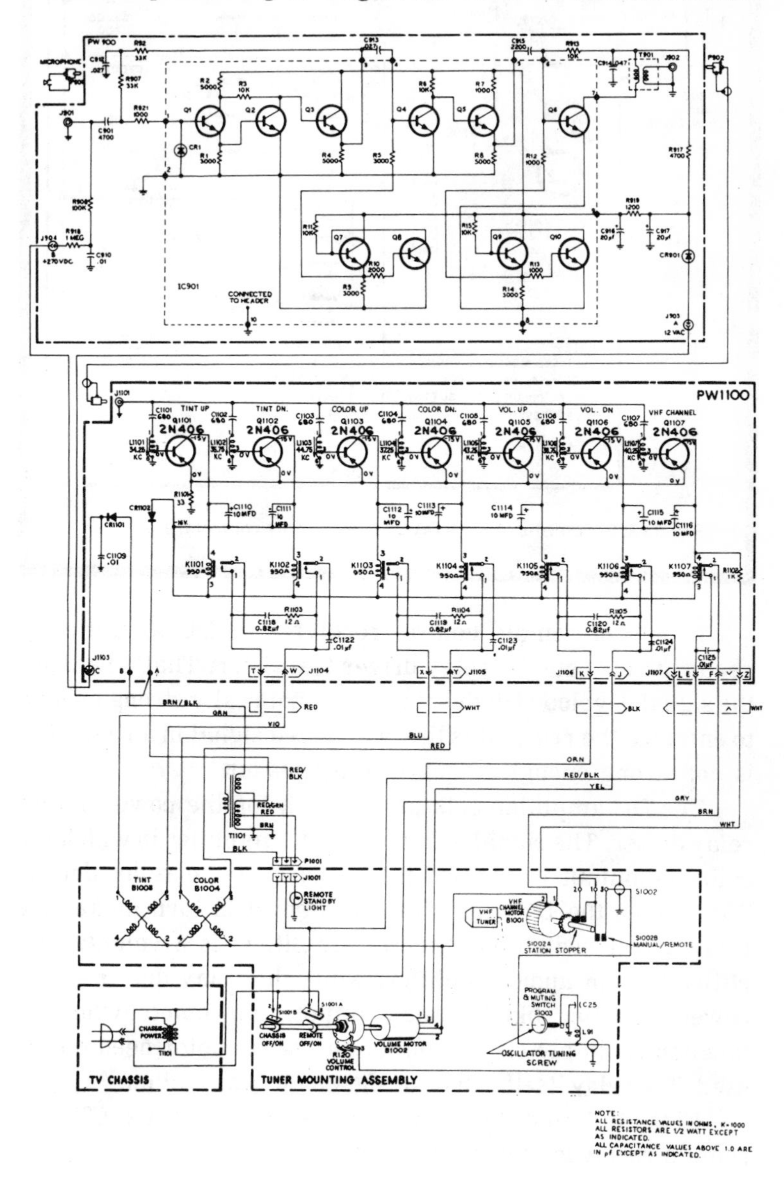

Fig. 13-5. Schematic of RCA's CTP11D remote control receiver.

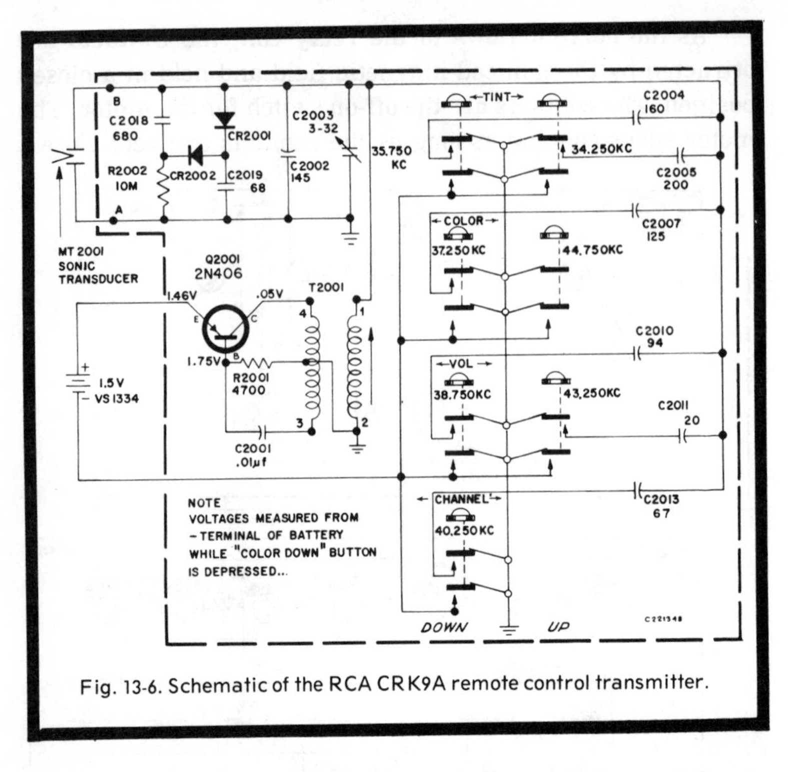

Fig. 13-6. Schematic of the RCA CRK9A remote control transmitter.

In lots of remote control receivers, a DC amplifier is needed to operate the relay driver transistor. That's because the signal developed in the receiver is not really strong enough to energize the relay. It is like a detected signal in a radio that is not strong enough to energize a speaker.

The DC amplifier acts as a switch for the power output relay driver. The signal is fed to the DC amplifier in which the collector output is coupled to the base of the relay driver. Then, when the DC amplifier turns on, it in turn makes the relay driver switch on. The DC amplifiers is a voltage amplifier like an audio amplifier, while the relay driver is the power amplifier, like the audio output. In a remote, when the function is simply off and on, of course, a motor need not be used. The relay itself can be designed for the function.

Figs. 13-5 and 13-6 are schematics of RCA's CTP11D remote receiver and transmitter.

CHAPTER 14
Separator Circuits

Separator circuits are distinguished from detection circuits because detection is synonynous with demodulation. In other words, a CW signal is modulated with information. The demodulation process extracts the information from the CW transmission.

Separation, as we refer to it here, occurs after demodulation. The intelligence that makes up modulation can be a single entity like amplitude modulation, frequency modulation or phase modulation. No separation of signals is needed. The intelligence is simply extracted and then fed to an output stage to perform its job. In some systems, though the intelligence contains a number of separate entities. When the carrier wave modulation contains a lot of different information units, each type of information must be separated from the rest and then processed on to do its work.

The composite color TV signal (Fig. 14-1) requires substantial amounts of separation. The composite signal information is contained in a 4.5-MHz range of the 6-MHz band. First there is the picture carrier. Around the picture carrier is the Y signal (the black-and-white range of video frequencies). The Y signal stretches out on the useful side of the carrier about 3.5 to 4 MHz. In the Y signal are all the whites, grays and blacks of the video. When the picture changes abruptly from white to black, the frequency at that instant is the maximum the TV can reproduce—between 3 and 4 MHz.

Next, riding piggy back on top of the Y signal range at 3.58 MHz, is the color subcarrier. Around the color subcarrier, extending for about a maximum of 1.5 MHz, are the color sidebands. At 4.5 MHz is the sound carrier. Around the sound carrier, for about 25 kHz, is the FM sound modulation.

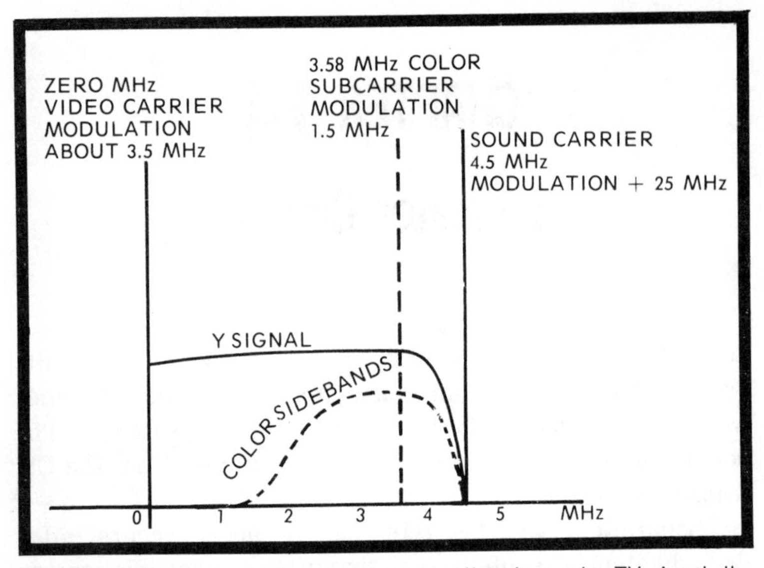

Fig. 14-1. Detection is not the same as separation. In a color TV signal, the detector processes the various carriers.

The TV picture is scanned from left to right. Each horizontal scan is followed by a horizontal sync pulse, which contains the scanning information. Located on the "back porch" of each sync pulse is the color burst of about eight cycles of a sine wave. This color burst synchronizes the color oscillator which converts the subcarrier (which is missing; that's why it's called a subcarrier) to an actual color carrier. Then at the end of every 263½ lines there is a vertical sync pulse so the picture can be scanned vertically at a 60-Hz rate (Fig. 14-2).

The separation circuits have to take apart all these individual signals so they may be processed alone to do their jobs. The Y signal is detected in a routine diode video detector. The sound is detected in a ratio detector type of circuit. The color side bands are picked out of the video with a tuned 3.58-MHz transformer. The horizontal sync, color burst and vertical sync are separated from the detected video.

SYNC SEPARATION

The sync pulses are contained in the flow of continuous signal from the transmitter. The video modulates the CW instant by instant. Each instant is a dot of light on the TV screen and the dot varies in brightness from white to black (Fig. 14-3). The dots are scanned from left to right, line by line, from top to bottom. The dots make up a line.

The video stops modulating at the end of each line. As soon as each line ends, a sync pulse is inserted. The sync pulse has a higher amplitude than the blackest dot. It is said to be in the "blacker than black" range. The amplitude is vital to the separation process, as well as blanking out the TV screen during the interval between the end of one line and the beginning of the next. The sync pulses occur at a 15,750-Hz rate. They are the "horizontal sync pulses," which eventually are fed to the horizontal oscillator to lock in step with the oscillator of the transmitter.

What about the vertical and color sync? They are also contained in the sync pulses. The vertical sync pulse is a result of a combination of horizontal sync pulses. Between every 262nd and 263rd horizontal pulse, the following happens:

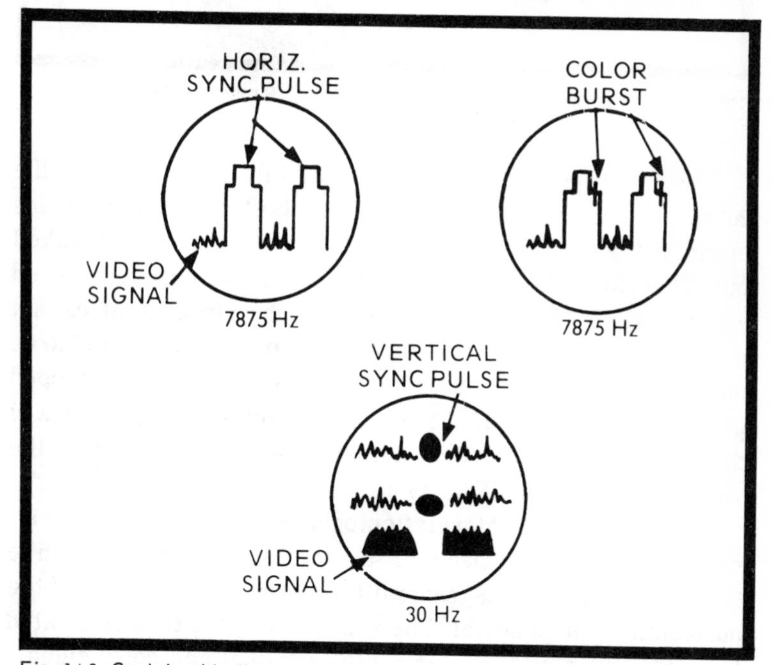

Fig. 14-2. Contained in the modulation are horizontal, vertical and color sync signals. These must be separated.

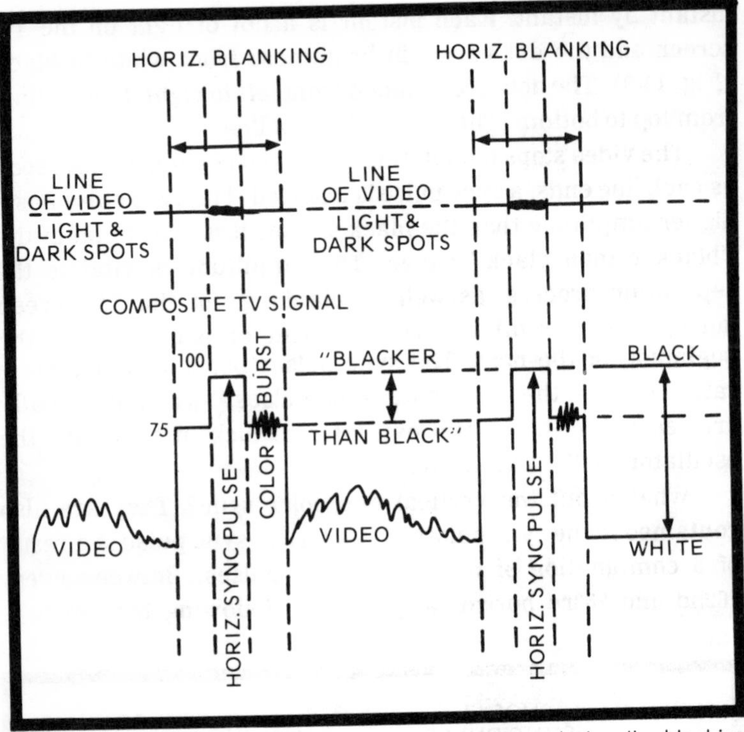

Fig. 14-3. Signals in the "blacker than black" region, during the blanking period between lines, have to be separated.

There is no video information at this time, because the 262nd pulse ends in the middle of the bottom line and begins again at the middle of the top line, and the screen is blanked out. Therefore, the top of the horizontal sync pulse is widened considerably. There is a capacitor in the circuit that cannot charge during the narrow-topped sync pulses, but can charge during the wide-topped pulses. The series of wide topped pulses occur at a 60-Hz rate. The capacitor's output (it will discharge as the next line starts) is 60 Hz and becomes the vertical sync pulse (Fig. 14-4).

The color sync information occurs at the same 15,750-Hz rate. A burst comprised of a sine wave with a definite frequency and phase is transmitted on the "back porch" of the horizontal sync pedestal. The sync pedestal is then separated from the video and the horizontal, vertical and color signals are extracted and fed to the respective oscillators to do their work; that is, locking the oscillator to the step of the transmitter.

SEPARATOR CIRCUIT

The separator circuit, also called a sync clipper, snips the entire sync pedestal off the composite signal. A typical sync circuit uses a PNP transistor in which the emitter-to-base junction is reverse biased and the transistor is cutoff. The signal is coupled to the base of the separator directly DC wise, but isolated with a resistor (Fig. 14-5). The value of the resistor is determined by the point in the video circuit where the sampling for the sync stage is tapped off. A good place to take the sample is at the video amplifier collector.

The signal that is coupled into the base must be negative-going. The base must be biased so that the video part of the signal is not negative enough to forward bias the stage. The separator is cutoff during the interval that the video arrives. However, when the even more negative sync pedestal reaches the separator, the bias becomes negative enough to forward bias the emitter-base junction and the transistor turns on during the sync interval. The collector-to-emitter current flows and in the collector an amplified and inverted sync pulse appears. The sync pulse that comes out of the separator is positive-going and has little or no distortion. The horizontal square, narrow pulses and the vertical square, wide pulses are intact. So the sync has been separated.

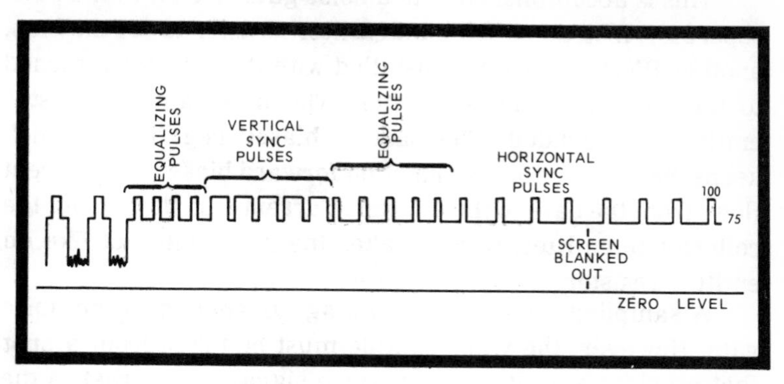

Fig. 14-4. The vertical sync pulses are simply a group of wider horizontal sync pulses.

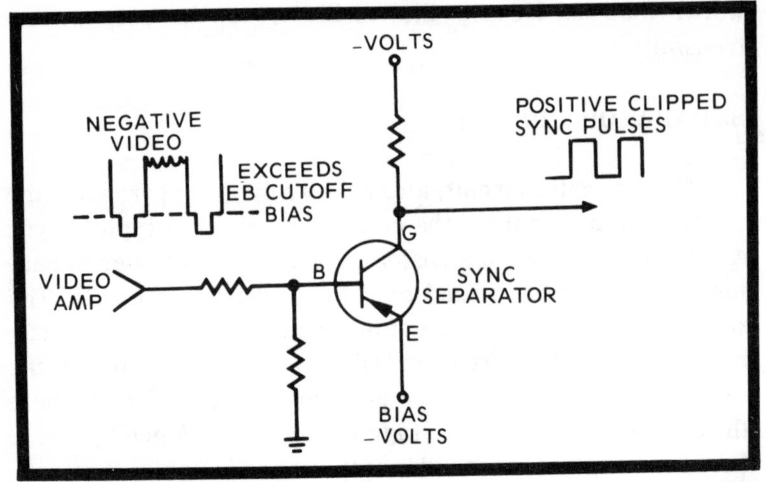

Fig. 14-5. A sync separator is biased at cutoff. When a pulse negative enough to exceed cutoff comes along, the separator conducts.

NOISE GATE

In most sync separators there is a noise gate circuit (Fig. 14-6). Noise pulses from outside interference appear on top of the video, often with a peak-to-peak height equal to the sync pulse. If a noise pulse should reach the base of the sync separator during the video interval, it can turn on the separator during the time it is supposed to be off. When this happens, the TV oscillators are triggered at the wrong time and the picture loses lock in; therefore, the noise pulses must not be allowed into the sync.

This is accomplished with a noise gate circuit. In the PNP separator, instead of using an emitter resistor to set the bias, another PNP transistor is installed with its collector attached to the emitter of the separator. The noise gate transistor emitter is grounded. The base is biased negatively, which keeps the emitter-to-base junction forward biased and current flows from the base to the emitter. Current also flows from the collector to emitter, which makes the noise gate act like an emitter resistor.

A sampling from the video stage is applied to the noise gate. However, the video sample must be taken from a spot that provides a positive-going video signal, in contrast to the negative signal in the separator. A diode is in series with the incoming signal. The diode is reverse biased by a voltage divider so that the diode conducts only when a high-level noise pulse appears. When the diode conducts, the positive noise pulse turns off the noise gate. This, in turn, causes the separator to stop conducting, since, effectively, the emitter of the separator is open with no gate conducting. The noise pulse that enters the separator is cancelled.

SYNC PULSE SEPARATION

While many TVs use complex vertical sync, horizontal sync etc., configurations, the separation of the horizontal sync from the vertical sync is accomplished with two simple circuits—a differentiator for the horizontal and an integrator for the vertical.

Differentiator

A differentiator circuit is nothing more than a high-pass filter (Fig. 14-7). It permits high-frequency pulses to pass

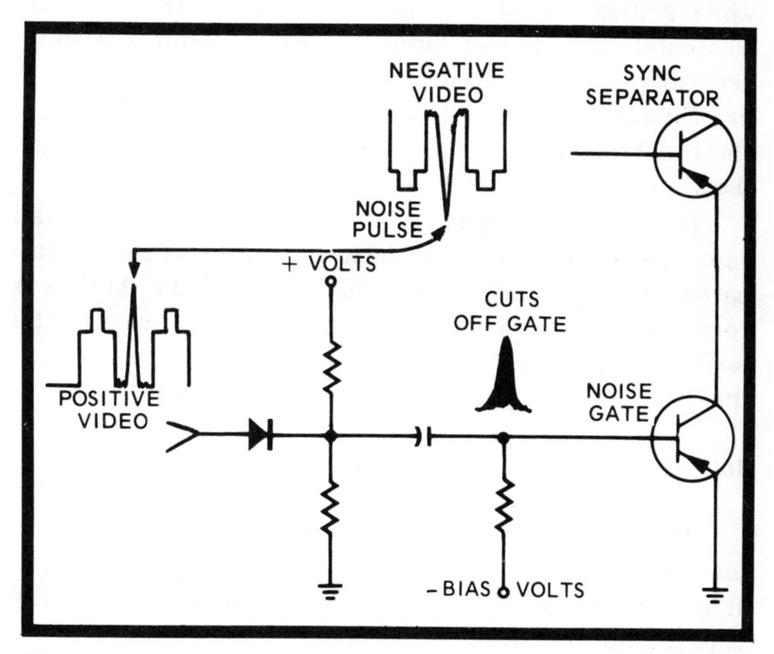

Fig. 14-6. A noise gate circuit is arranged to cut off the sync separator during any noise pulses.

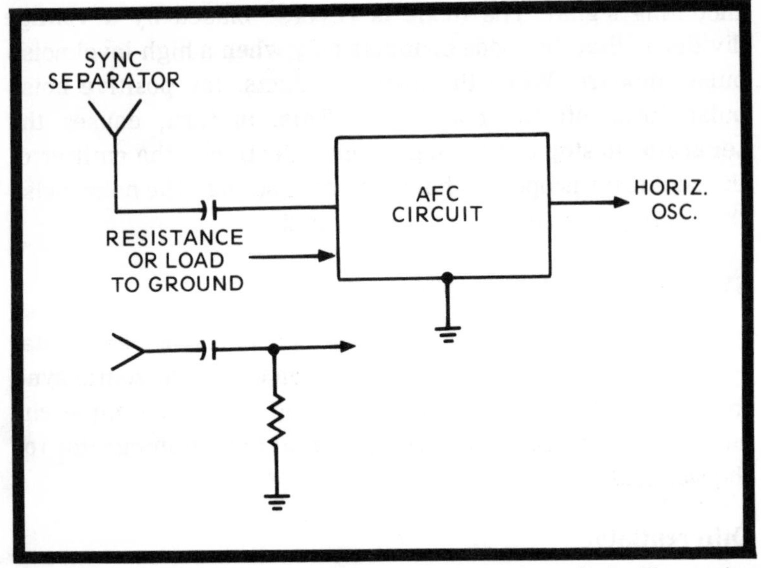

Fig. 14-7. A differentiator circuit shapes the separator output with a series capacitor and resistance to ground.

while getting rid of any low frequencies. The 15,750-Hz horizontal pulse passes through it easily, while the 60-Hz vertical pulse is stopped. The circuit consists of a small capacitance in series with a resistance to ground. Typically, the resistance turns out to be an AFC circuit, such as the discriminator type (see Chapters 11 and 12).

The horizontal sync pulse passes through the small capacitance and enters the phase detector where it is compared to the flyback pulse. The output of the AFC is a correction voltage that controls the frequency of the horizontal oscillator. The correction voltage varies as a direct result of the frequency and phase of the horizontal pulse.

Integrator

An integrator is a low-pass filter (Fig. 14-8). It permits low-frequency pulses to pass and blocks any high frequencies. It is composed of a resistor in series and a medium size capacitor to ground. The output of the integrator, which is a 60-Hz pulse, is passed right into the vertical oscillator.

COLOR SYNC SEPARATOR

The color sync is also present on the horizontal sync pedestal. It is about eight sine-wave cycles behind the sync pulse on the "back porch." The sine waves maintain the correct frequency and phase of the color carrier.

The color carrier is suppressed at the transmitter. The only thing transmitted is the color modulation. Therefore, the color carrier must be reconstructed in the color receiver. After reconstruction, it is then remodulated with the color sidebands and further processed. The rest of the processing consists of detection and then separation of one color from another for final display.

The color carrier is reconstructed by an oscillator in the receiver. The oscillator is crystal controlled and runs near the prescribed frequency of 3.58 MHz. However, it runs free. It is not exactly on 3.58 MHz nor exactly in phase with the transmitter. The oscillator has to be triggered so it is in frequency and phase. The information to do this job is contained in the color burst. The burst must be extracted from the sync pulse back porch and passed to an AFC phase detector. In the phase detector the burst is compared to the free-running oscillator's frequency and phase. A correction voltage is produced and fed to a reactance circuit or a varactor diode to keep the oscillator in perfect timing with the transmitter.

The burst amplifier is typically held cutoff with a reverse bias across the emitter-to-base junctions. The sampling of the video is applied to the base. The video cannot enter during cutoff. A pulse from the flyback transformer is fed to the base

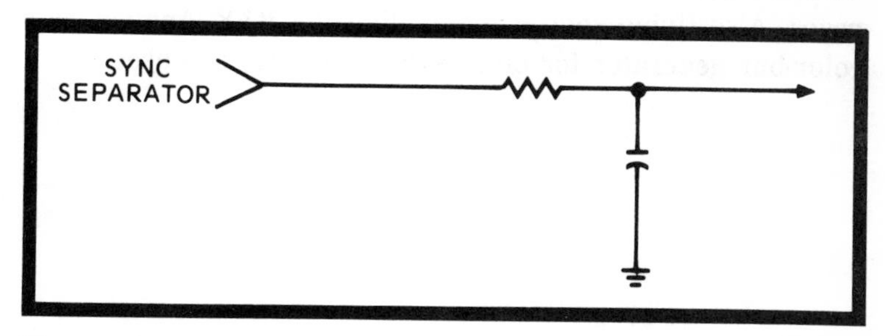

Fig. 14-8. An integrator circuit shapes the separator output with a series resistor and capacitor to ground.

or emitter. The pulse is either negative or positive according to its point of entry in the transistor, but it must be the correct polarization to key on the device. The device conducts during the time that the flyback pulse arrives. The pulse is there during the time the burst is arriving at the base. The burst appears amplified in the collector of the burst amplifier, then it is coupled through a transformer to the color phase detector. The burst amplifier is quite simple. It is keyed to separate the burst signal.

TESTING SEPARATION CIRCUITS

The best way to test separator circuits is with a scope equipped with a high-impedance probe. The frequencies are all viewable since the circuits are all located after the video detector. Most service notes provide little sketches or actual photos of the correct peak-to-peak waveform. Testing consists of looking closely at the scope pattern and comparing them with the recommended waveform. A good VTVM can also give the peak-to-peak values.

Any deviation or distortion of the prescribed peak-to-peak waveforms is an indication of trouble. Certain waveforms should appear at specific test points; if one is not there, it is a clear indication of trouble. The signal can be followed from component to component. If it suddently disappears, the trouble is near.

Separator circuit troubles are among the most confusing types. There is no easy signal injection method; the transmitted signal provides the best type of signal for test purposes. Also flying spot scanners like the B&K Analyst or a color-bar generator for color sync problems are useful.

CHAPTER 15

Power Supplies

A transistor power supply does not provide heater voltages, of course. But other than that simplification, the transistor supply is at least as complex as a tube supply and usually more so.

HALF-WAVE

The conventional half-wave rectifier circuit is well known. A rectifier of the proper current rating, usually able to pass 500 to 1000 ma, is installed in series with the line (Fig. 15-1). The cathode of the rectifier goes to the B+ connection, while the anode connects to the AC input.

On the positive half cycle of the 60-Hz AC wave, the anode attracts electrons from the receiving B+ lines. The rectifier is forward biased and electrons flow from B+ to B minus or ground. This is opposite to the way electrons flow in the receiver (or transmitter). In the receiver, electrons flow from B minus to the B+. In an NPN, the electrons come from ground, pass through the NPN from emitter to collector and then go into the power supply.

In a PNP, the electrons also come from ground, but pass from the collector to the emitter and then into the power supply. It really doesn't matter which way the electrons flow in the transistor, collector current is still effective collector current. The DC collector current can still develop the peak-to-peak signal across a load no matter which direction the electrons are flowing. With NPNs, PNPs, N-channel or P-channel FETs, the electrons in the power supply still pass from B+ through the forward biased rectifier to B minus.

On the negative half cycle of the AC input, the rectifier develops a negative charge on the anode. This reverse biases the rectifier and no current flows during the negative half

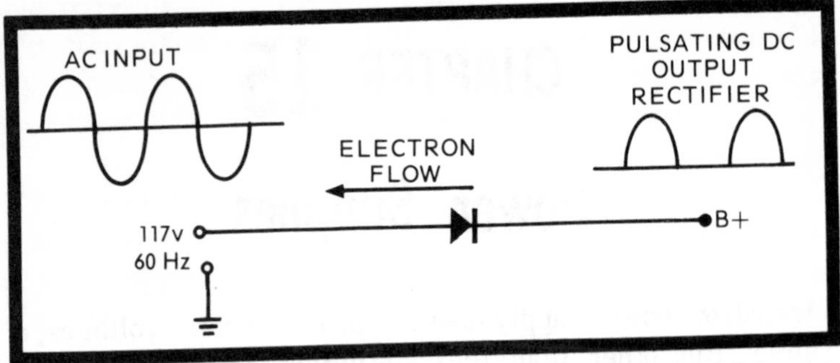

Fig. 15-1. A single rectifier allows the positive pulses from the AC input to appear in the output. The negative half cycles of the input are blocked.

cycle. In other words, electrons can pass only in one direction due to the rectifier. The AC is changed to a pulsating DC.

FULL-WAVE BRIDGE

In tube power supplies a dual-diode is used to produce full-wave rectification instead of half-wave. The full-wave is more efficient than the half-wave, since it uses the negative half cycle of energy instead of simply discarding it like the half-wave does.

In transistor power supplies, the full-wave also can be utilized with two diodes, but more than likely four diodes are used in a bridge rectifier configuration (Fig. 15-2). A power transformer is usually employed and the four diodes are connected to the secondary of the transformer. (The transformer doesn't have to be used; the bridge could also be attached directly to the line.) The transformer isolates the receiver from the line, eliminating the shock hazard of having one side of the line attached directly to the chassis.

The diodes are wired in two series pairs and the two pairs are in parallel. Two cathodes are tied together and connected to B+. While two anodes are tied together and grounded. The secondary of the transformer is connected to the two intersections in the center of the series pairs.

When the receiver is turned on, AC enters the transformer. As the positive half of the cycle appears at the top of the secondary, the negative half appears at the bottom. This puts a positive voltage on the anode of the left top rectifier. It draws electrons from the B+ line due to its forward bias. Simultaneously, a negative voltage is placed on the cathode of the right bottom rectifier. Since its anode is grounded, this rectifier becomes forward biased. Electrons pass from cathode to anode and return to ground.

As the negative half of the cycle appears at the top to the secondary, a negative charge is placed on the cathode of the first rectifier. The positive half of the cycle appears at the bottom of the secondary, placing a positive charge on the anode of the second rectifier. Both the right top and left bottom rectifiers are forward biased and conduct. The advantage of using the bridge setup instead of just two rectifiers for full-wave operation is that the peak inverse load is distributed over two rectifiers instead of one. The PIV ratings of the rectifiers need only be half as large. The waveshapes are identical.

ACTIVE POWER FILTERS

Filter capacitors, chokes and combinations of these items are needed to smooth out the pulsations in the DC rectifier output. An ordinary filter is a large electron storage device. As the rectifier conducts, the filter charges to the peak voltage of the rectifier output. Then as the cycle approaches the zero

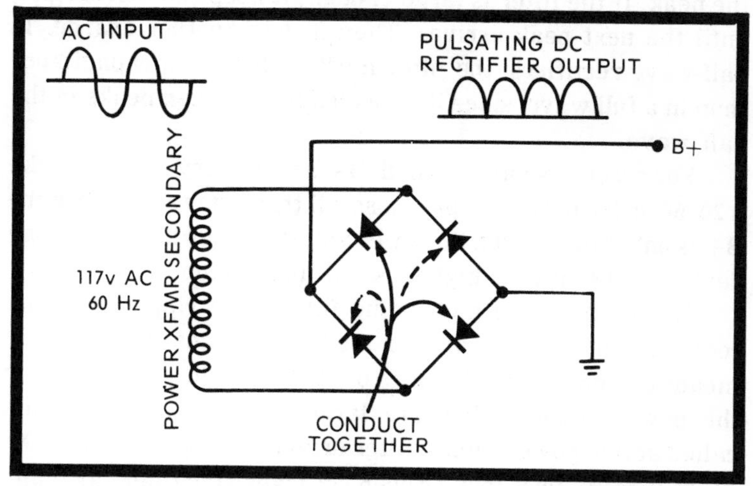

Fig. 15-2. A full-wave bridge rectifier distributes the circuit load so PIV ratings need not be so high.

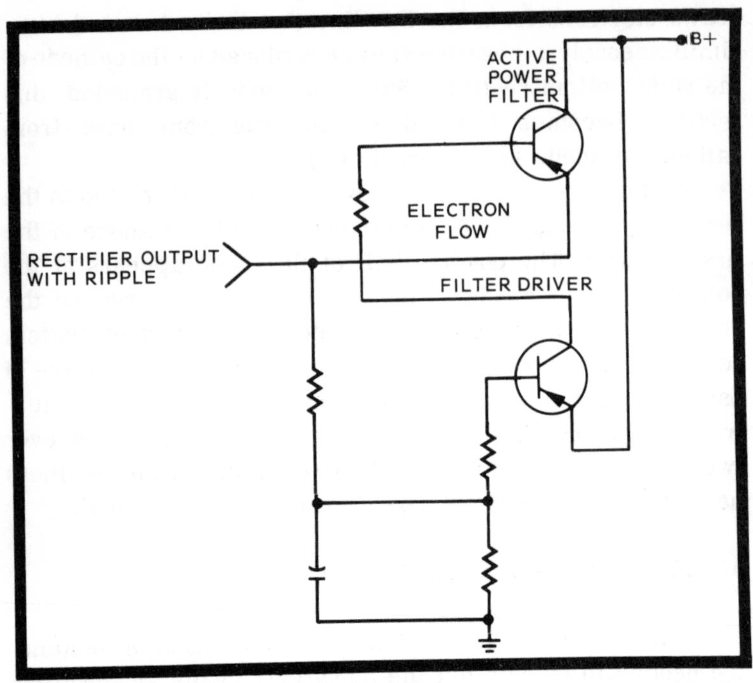

Fig. 15-3. A transistorized active power filter and driver get rid of ripple by the noise cancellation method.

base line, the filter discharges, keeping the actual output at the peak. If the filter is large enough, it keeps on discharging until the next peak arrives. Then it gets another charge. In half-wave rectifiers, the filter must be larger in capacitance than in a full wave, since it takes longer between peaks in the half wave.

For receivers having a high B+ an ordinary size filter like a 20, 50 or 100 mfd is ample. In small transistor sets where the B+ is only 9 or 12 volts, the lower voltage needs higher storage devices; 1000-mfd or higher is not uncommon.

Even with these high value filters, the ripple from the rectifier is hard to get rid of in the low B+ lines. So electronic means of ripple cancellation is used. Transistors are placed in the power supply. They are large output type transistors called active power filters, filter drivers and so forth (Fig. 15-3). Typically, two, four or even six transistor filters are common. They operate in a noise cancellation manner and also remove most low frequencies including the 60-Hz vertical sweep and 15,750-Hz horizontal sweep rates that might be coupled into the power supply. These frequencies, if not filtered, will be distributed to the rest of the receiver and cause all types of interference and distortion.

As the electrons are drawn from the circuits into the power supply, they encounter two PNP active power filter transistors. The filter collector is attached to the B+ line and the emitter connects to the rectifiers. Electrons flow from B+ to the collector, through the PNP, out the emitter to the rectifiers. Any ripple passes through the PNP along with the current flow.

The filter driver is connected the same way. Another PNP emitter goes to the B+ line and its collector is attached to the base of the other PNP. The filter driver base is hooked to ground through a voltage divider. At the center of the voltage divider is a sampling of the top PNPs emitter output, isolated from the top stage with a resistor. Any ripple in the current is developed across the emitter-base junction.

The driver amplifies and inverts the ripple, then injects it into the base of the active power filter. The amplified and inverted ripple from the driver enters the power filter exactly out of phase with the ripple that is passing through. They cancel each other. In fact, they cancel each other to such an efficient degree that, for all intents, the ripple is gone.

HIGH-VOLTAGE POWER SUPPLIES

A high-voltage power supply in transistor circuits is mainly used to provide a high DC picture tube anode voltage, just as in a tube circuit. The high voltage is produced by the sudden collapse of the flyback pulse during the horizontal scan retrace time.

When the pulse collapses, the ensuing magnetic field that is developed, due to the suddenness of the current movement, is very high. This large magnetic field occurs in the primary of an autotransformer hookup in the flyback transformer. The primary passes the magnetic field into the secondary where another quick current is developed and the voltage is stepped up. At the top of the transformer the voltage is taken off and rectified. Depending on design, the voltage can be as high as 30 kilovolts.

The rectifier circuit is a simple half-wave type; the only difficulty is the large amount of voltage involved. Insulation quality is one of the most important considerations. Most failures are due to the circuit not being able to handle and contain the high voltage. Motorola has developed a high-voltage rectifier that is made of solid-state materials. This circuit is exactly like its tube counterpart, except for the solid-state rectifier instead of the old tubes.

VOLTAGE MULTIPLICATION

Another approach to the high-voltage supply situation uses a transformer that produces only a fraction of the needed voltage. For instance, Zenith has a flyback in a color chassis that produces only a little over 8,000 volts. The 8 KV is then put into a special solid-state printed network that ripples the available voltage. Some TVs have voltage doublers and others voltage quadruplers. How can the high voltage be multiplied?

It's easy and an old technique. Sylvania had a voltage doubler way back in 1950 not unlike the ones used today. The only difference is the multiplier rectifier units today are solid state. Voltage is the electrical pressure that pushes the electrons along. Picture tube anode current is relatively small, only about 1 ma. The current remains the same, only the voltage is multiplied. Also, the flyback frequency is 15,750 Hz. In a 60-Hz circuit, voltage multiplication is a problem. Filter capacitors must store a charge for a longer time between pulses, otherwise voltage regulation is poor.

In the high-voltage system, the higher sweep frequency allows smaller values of capacitance to be used. In a voltage multiplier there are several diodes and capacitors. In Fig. 15-4, as the first flyback pulse arrives at the input, D1 is forward biased and electrons are drawn from C1, giving it a peak charge of about 8 KV. Then as the pulse falls off, the charge on C1 forward biases D2, pulling electrons out of C2. C2 charges and the flow through D2 reduces the charge on C1. In effect, C1 and C2 share the original 8 KV charge, giving both of them a 4 KV charge.

Then, the next pulse arrives and again forward biases D1. This pulls a lot of electrons out of C1 again, recharging it to 8 KV. Then the pulse recedes. D2 again conducts, since C1 has 8 KV while C2 only has 4 KV. D2 is thus forward biased. The two capacitors again share their total, which is 8 KV on C1 and 4 KV on C2. The total of 12 KV shared equally puts 6 KV on both C1 and C2.

When the third pulse arrives, the 8 KV on the anode of D1 forward biases D1, pulling more electrons from C1. This allows C1 to accept an even higher charge of 6 KV plus the new 8 KV, making 14 KV. As the pulse recedes, C1 again shares its charge with C2. They now each have 7 KV. This process continues for a few more pulses until C1 and C2 both come up to full charge of 8 KV (actually it is 8.3 KV).

Once the voltage at C1 is at full charge, the next input pulse of 8 KV can't draw any more electrons out of C1. The pulse adds to the C1 charge, forward biasing D2. D2 can't draw any more electrons out of C2. The pulse adds to the C2 charge, making the accumulated charge 16 KV. This forward biases D3.

D3 draws electrons out of C3. This forward biases D4. D4 draws electrons out of C4. D3, D4, C3 and C4 repeat the same charge and discharge procedure that D1, D2, C1 and C2 performed. The only difference is they charge up to 16 KV. This is the 8 KV charge of C1-C2 plus the next 8 KV charge of C3-C4. They keep charging until all the possible electrons are drawn

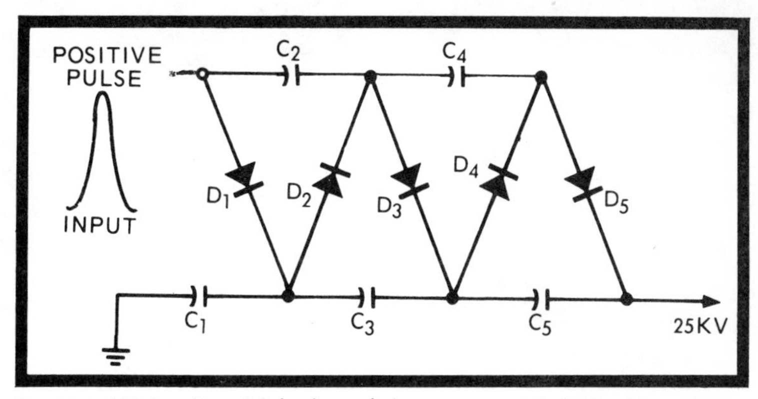

Fig. 15-4. A high-voltage tripler is made in an encapsulated PC unit containing five capacitors and five diodes.

from the capacitors to create the high positive charge of 16 KV (actually it is 16.6 KV).

Once the charge at the junction of C4 and D4 reaches 16 KV, D5 becomes forward biased. Electrons are drawn out of C5. The pulses of 8 KV each cause C5 to also charge due to a lack of electrons. The voltage has been rectified, filtered and tripled. Three times 8.3 KV is approximately 25 KV, the desired CRT DC anode potential.

Solid-state voltage triplers are small plastic encapsulated devices. They come in various shapes but are all basically alike.

INDEX

A

AC beta, transistor, 42
Acceptor material, 11
Active power filters, 215
Adjacent-channel sound carrier, TV, 106
Adjacent-channel video, 106
AFC, 184
AGC, 178
AGC, IF, 93, 98
AGC, RF, 80
Alignment, discriminator, 174
Alignment, IF, 98
Amplification, transistor, 42
Audio amplifiers, 110
Audio output transformer, 129
Automatic frequency control, 184
Automatic gain control, 178
Automatic gain control, RF, 80

B

Balun transformer, 76
Bandpass, IF, 89
Bandpass, video amplifier, 117
Bandwidth, FM, 175
Barrier region, 18
Base, transistor, 33
Beta, transistor, 42
Bias, forward, 20, 21
Bias, IF alignment, 109
Bias, reverse, 20
Bias, SCR, 64
Bias, transistor, 32, 33
Bipolar transistor, 31
Breakover voltage, SCR, 66
Bridge mixer, 163
Bridge rectifier, 214
Brightness control, 119

C

Capacitive AGC control, 188
Capacitor, IC, 61
Cascade RF amplifier, 83
Channel, FET, 46
Class A amplifier, 111
Class B amplifier, 111
Clipper, sync, 207
Collector, transistor, 33
Color AFC, 184
Color subcarrier, 106
Color sync separator, 211
Common-emitter RF amplifier, 73
Composite color TV signal, 203
Conductor, 7
Contrast control, 119
Conversion loss, 157
Converter tests, 164
Converter, transistor, 157
Converter, tube-type, 156
Coupling, IF transformer, 90
Coupling transformer, transistor receiver, 166
Covalent bonds, 7
Critical coupling, IF transformer, 92
Crystal oscillator, 145
Crystal structure, 10
Current flow, N material, 12
Current flow, P material, 15

D

DC beta, transistor, 42
Deflection yoke, 129
Delay line, 118
Depletion mode, FET, 50
Depletion mode IGFETS, 51
Detector load, 111

Detector diodes, 25
Dielectric, IGFET, 51
Differentiator, horizontal, sync, 209
Diode, 17
Diode capacitance, 26
Diode detector, 167
Diode mixer, 161
Diode test, 21, 39
Direct conversion, 163
Discriminator alignment, 174
Discriminator, FM, 170
Donor material, 12
Drain current, FET, 55
Drain, FET, 45
Dual-gate FET mixer, 161

E

Electron flow, JFET, 47
Electron flow, N material, 12
Emitter, transistor, 33
Enhancement mode, FET, 51
Enhancement mode, IGFETS, 51

F

Feedback, LC oscillator, 149
FET mixer, 159
FET RF amplifier, 83
FETs, 45
Field-effect transistors, 45
Filter, active power, 215
Flyback transformer, 131
FM detectors, 170
Forward bias, 20, 21
Forward bias, RF amplifier, 81
Forward bias, SCR, 64
Forward bias, transistor, 36
Forward breakover voltage, SCR, 66
Frequency control, oscillator, 145
Frequency, oscillator, 154
Full-wave rectifier, 214

G

Gain FET, 54
Gain, RF amplifier, 70
Gate, FET, 45
Gate leakage, FET, 54
Gated AGC, 178
Germanium, 12
Germanium diode, 25

H

"Hale" flow, 14
Half-wave rectifier, 213
Heat sensitivity, crystal, 147
High-frequency compensation, video amplifier, 116
High-frequency response, video amplifier, 121
High-voltage power supply, 217
Homodyne converter, 164
Horizontal AFC, 184
Horizontal output, 130
Horizontal output circuits, 134

I

Icbo, transistor, 42
IC production, 61
ICs, 59
IC tests, 62
IF AGC, 93, 98
IF alignment, 98
IF amplifier, 88
IF transformers, 90
IF traps, TV IF, 102
IG FET, 50
IG FET handling, 56
IGSS, FET, 54, 55
Impurities, 10
Inductance, IC, 59
Inductive AFC control, 187
Inert state, 8
Input circuit, RF amplifier, 74
Insulated gate FET, 50
Integrated circuits, 59
Integrator, vertical sync, 210
Intermediate amplifier, 88
Internal noise, RF amplifier, 71
Isolation, RF amplifier, 72

J

JFET, 47
JFET mixer, 159

Junction bias, 19
Junction FET, 47

K

"Knee" voltage, zener, 29

L

LC oscillator, 149
Leakage, transistor, 43
Linear amplification, RF amplifier, 72
Limiter, 171
Loading, IF transformer, 93
Low-frequency compensation, video amplifier, 115
Low-frequency response, video amplifier, 121

M

Majority carriers, 12
Marker injection, 108
Metal Oxide Semiconductor FET, 50
Minority carriers, 16
Mixer, FET, 159
Mixer tests, 165
Mixer, transistor, 158
MOSFET, 50
Multiplier, 148
Multivibrator, 152

N

N-channel FET, 46
N-channel IGFET, 51
Neutralization, 79
Neutralization, IF, 95
N material, 12
Noise gate, 208
Noise, RF amplifier, 70

O

Oscillator frequency, 145
Oscillator tests, 154
Oscilloscope connections, IF alignment, 103
Output circuit, RF amplifier, 78
Output transformer, 128

P

P-channel FET, 47
P-channel IGFET, 51
Peaking coil, 116
Peak inverse voltage, 24
Pentavalent material, 11
Piezoelectric effect, 146
Pinchoff, FET, 49
P material, 14
PN junction, 17
Post injection, marker, 108
Power amplifiers, 128
Power output tests, 136
Power rectifiers, 24
Preinjection, marker, 108
Pushpull amplifier, 131

Q

Q, IF transformers, 90
Quartz crystal, 146
Q, varicap, 31

R

Radio IF alignment, 101
Ratio detector, 175
RC oscillator, 151
Receiver, RF remote control, 199
Receiver, supersonic remote, 194
Relay driver, remote control, 199
Remote control receiver, 199
Remote control, RF, 192
Remote control, supersonic, 193
Resistance, diode junction, 22
Resistance, IC, 61
Reverse bias, 20
Reverse bias, RF amplifier, 81
Reverse bias, transistor, 36
Reverse current, germanium diode, 26
Reverse current, silicon diode, 24
RF AGC, 80

RF amplifier, 70
RF amplifier tests, 84
RF power amplifier, 135
RF remote control, 192
RF remote control receiver, 199
RF transmitter, remote control, 193

S

SCRs, 64
"S" curve, 175
Selectivity, RF amplifier, 72
Semiconductor, 7
Separator, sync, 204
Signal injection, IF, 98
Signal injection, video amplifier, 121
Signal tracing, IF, 98
Signal tracing, video amplifier, 121
Silicon atom, 7
Silicon controlled rectifiers, 64
Silicon diode, 24
Sound carrier, 106
Source, FET, 45
Speaker, 129
Stabilization, IF, 95
Stagger tuning, IF, 89
Substrate, FET, 46
Supersonic remote control, 193
Supersonic remote control receiver, 194
Supersonic transmitter, 194
Sweep generator, IF alignment, 103
Synchrodyne converter, 164
Sync separation, 204

T

Test, diode, 40
Test, IC, 62
Testing, diode, 21
Test pattern response, 121
Test, transistor, 41
Thyristor, 66
Time constant, diode detector load, 168
Transducer remote control, 194
Transducer transmitter, 194
Transformerless pushpull output, 132
Transformers, IF, 90
Transistor converter, 157
Transistor IF, 90
Transistor mixer, 158
Transistor test, 41
Traps, 97
Traps, TV IF, 102
Trivalent atom, 10
Tube-type converter, 156
Tuned input, RF amplifier, 76
Turn-on voltage, 26
Turn-on voltage, silicon diode, 24
TV IF alignment, 102
TV tuner alignment, 85

V

Valence shell, 9
Varactor diodes, 188
Varicap diode, 30
Varicap tests, 40
Vertical output circuits, 133
Vertical output transformer, 129
Video amplifier, 115
Video IF frequency, 106
Voltage amplifier, 113
Voltage multiplication, high voltage, 218
Volume control, 111

Y

Yoke, deflection, 129

Z

Zener diode, 28
Zener diode checks, 40